D0385491

BEYOND DUMPING

New Titles from QUORUM BOOKS

Telecommunications America: Markets Without Boundaries
MANLEY RUTHERFORD IRWIN

Pension Fund Investments in Real Estate: A Guide for Plan Sponsors and Real Estate Professionals
NATALIE A. McKELVY

Growth Industries in the 1980s: Conference Proceedings
FEDERAL RESERVE BANK OF ATLANTA, SPONSOR

Business Strategy for the Political Arena
FRANK SHIPPER AND MARIANNE M. JENNINGS

Socio-Economic Accounting
AHMED BELKAOUI

Corporate Spin-Offs: Strategy for the 1980s
RONALD J. KUDLA AND THOMAS H. McINISH

Disaster Management: Warning Response and Community Relocation
RONALD W. PERRY AND ALVIN H. MUSHKATEL

The Savings and Loan Industry: Current Problems and Possible Solutions
WALTER J. WOERHEIDE

Mechatronics
MICK McLEAN, EDITOR

Employee Counseling and Employee Assistance Programs
DONALD W. MYERS

The Adversary Economy: Business Responses to Changing Government Requirements
ALFRED A. MARCUS

Microeconomic Concepts for Attorneys: A Reference Guide
WAYNE C. CURTIS

BEYOND DUMPING

New Strategies for Controlling Toxic Contamination

edited by
BRUCE PIASECKI

Q
Quorum Books
Westport, Connecticut • London, England

Library of Congress Cataloging in Publication Data

Main entry under title:

Beyond dumping.

Bibliography: p.
Includes index.
1. Hazardous wastes—Law and legislation—United States. 2. Hazardous waste sites—Law and legislation—United States. I. Piasecki, Bruce.
KF3958.B49 1984 344.73′0462 83-24510
ISBN 0-89930-056-1 (lib. bdg.) 347.304462

Library of Congress Catalog Card Number: 83-24510
ISBN: 0-89930-056-1

First published in 1984 by Quorum Books

Greenwood Press
A division of Congressional Information Service, Inc.
88 Post Road West, Westport, Connecticut 06881

Printed in the United States of America

10 9 8 7 6 5 4 3 2 1

Contents

PART II: STATE AND EUROPEAN INITIATIVES FOR STRONGER CONTROLS

PART III: NEW GROUNDS FOR HOPE

Foreword

In 1976, Congress enacted the Resource Conservation and Recovery Act (RCRA) as the nation's basic authority governing the management of hazardous wastes. RCRA established an ambitious "cradle to grave" regulatory approach to hazardous wastes, making it possible to track the existence and the ultimate disposal of those wastes. At the time of RCRA's enactment, Congress and the American people knew they were facing significant environmental problems but were unaware of the extent of those problems. The efforts of the Environmental Protection Agency (EPA) to implement RCRA have demonstrated that the task of comprehensive hazardous waste management is one of unparalleled scope and complexity.

Under RCRA, EPA was given 18 months to issue regulations governing the identification, generation, transportation, storage, treatment, and disposal of hazardous wastes. The agency, however, did not publish its first major package of regulations until May 1980. As disappointing as this delay was, these requirements nevertheless represented an important first step toward the responsible management of the nation's hazardous wastes. Since May 1980, EPA's promulgation of several proposed final standards for treatment, storage, and disposal facilities and financial responsibility requirements has at least created the basic framework of a hazardous waste regulatory and enforcement program.

We in Congress have learned a great deal since 1976 about the nation's hazardous waste problems. In our oversight of EPA's implementation of RCRA, we have discovered problems that were either unknown in 1976 or only vaguely perceived. Chief among these problems is the number of abandoned hazardous waste sites scattered throughout the country. Beginning with the Love Canal tragedy in New York, each day seemed to bring new revelations about the problem. In response to this crisis, Congress eventually enacted the Comprehensive Environmental Response, Compensation, and Liability Act of 1980 (CERCLA), now better known as Superfund.

CERCLA created a $1.6 billion fund to clean up the nation's abandoned hazardous waste sites. Yet, as Love Canal recedes into memory and new names, such as Times Beach, Missouri, replace it in the nation's lexicon of hazardous waste disasters, we have discovered that $1.6 billion is totally inadequate to clean up these sites. EPA has published a list of 419 sites which have been declared priority Superfund sites. However, the $1.6 billion Superfund will only be able to clean up about 45 of these priority sites. Although no one knows for certain, it has been recently estimated that there are now as many as 16,000 abandoned hazardous waste sites in this country and that it could cost $40 billion or more to clean them all up.

Clearly, then, this country simply cannot afford the irresponsible disposal of hazardous wastes. According to EPA, the cost of cleaning up improperly disposed waste is approximately $2,000 per ton, while the cost of disposing of waste in accordance with RCRA is only $90 per ton. For purely monetary reasons, RCRA's regulatory and enforcement programs make far more sense than allowing new Superfund sites to develop which will have to be cleaned up years from now at a much higher price.

Yet, more important than the cost of uncontrolled disposal is the danger to public health and the environment. Improperly disposed hazardous wastes threaten both the air and surface water, but the threat to groundwater is especially acute. Americans rely heavily on groundwater for their drinking water, often using wells without any treatment equipment whatsoever. A recent EPA study of underground drinking water supplies in 954 cities showed contaminated water in almost 300 municipalities, each with a population over 10,000. This statistic is due in large measure to the fact that many land disposal facilities do not comply with current law. Another EPA study found that 109 of the 171 disposal sites studied were violating RCRA regulations.

Yet setting standards for landfills and enforcing those standards is not the ultimate answer. Our experience since 1976 has shown us that there is no such thing as a secure landfill. We must finally realize that the land disposal of hazardous wastes cannot continue. As noted above, it is land disposal which threatens a large portion of our nation's drinking water supply. Congressional hearings have revealed that alternatives to land disposal exist right now and are underutilized. Land disposal, therefore, must become the method of last resort. Once other methods of disposal are fully utilized, landfills will become the repositories for only those wastes which are otherwise untreatable.

In the search for alternatives to land disposal, the federal government can profit from the experience of several states. It has often been true that the states have supplied a laboratory for testing new ideas. In the area of hazardous waste disposal, California, for example, has been in the forefront of the search for alternatives to land disposal. Other states have also established stronger controls over the land disposal of hazardous wastes, realizing that this is the least desirable method.

According to recent studies, there are approximately 245 million tons of hazardous waste generated in this country each year. This is more than 2,000 pounds for every man, woman, and child in the United States. Yet only about 40 million tons of this waste are subject to controls under the present federal system. Since much of this regulated hazardous waste is being dumped into landfills, it is inevitable that the nation's drinking water supply will remain threatened and new Superfund sites will be created.

To even the most casual observer, it must be clear that there are loopholes in the present federal system. As I write this foreword in the early summer of 1983, I am working for the passage of H.R. 2867, a bill which I introduced into the House of Representatives not only to reauthorize RCRA for fiscal years 1984 through 1986, but also to close the gaps in the present regulatory and enforcement program. For example, H.R. 2867 will severely limit the disposal of containerized liquid hazardous wastes in landfills and prohibit the landfill disposal of bulk or non-containerized liquid hazardous wastes.

It will also ban outright from land disposal certain specified wastes, including arsenic, mercury, and cyanide. In addition, H.R. 2867 will require regulation of hazardous wastes for small-quantity generators (defined as those who generate a total quantity of between 100 and 1,000 kilograms of hazardous waste per calendar month). The Office of Technology Assessment has estimated that each year as much as 4 million tons of hazardous waste are disposed of in an environmentally unsound manner because small-quantity generators are not subject to regulation.

On the enforcement side, H.R. 2867 provides for stiffer criminal penalties for those who deliberately violate RCRA and grants EPA investigators law enforcement powers to aid them in their investigations into criminal activity. These and the bill's other enforcement provisions are necessary because much of the illegal dumping of hazardous wastes in this country is not due to mere ignorance of the law, but to the increasing practice of "midnight dumping." Tougher enforcement including fines and prison sentences provide the best means for plugging this particular loophole—the source of many of our current Superfund sites.

This leads to my final point, which is the critical role EPA plays in the hazardous waste area and its recent flawed performance. It is axiomatic that regulatory statutes passed by Congress will have little or no impact unless the responsible executive branch agency issues the necessary regulations and then enforces those regulations. As mentioned above, EPA has been extremely slow in issuing the required RCRA regulations. The more disturbing aspect of EPA's performance during the past two years has been the lack of commitment on the part of top EPA officials to the basic principles underlying the nation's environmental legislation and a distinct lack of enthusiasm for enforcing those laws, especially RCRA and Superfund.

I hope this is all in the past. Ironically, EPA's tribulations have pointed out certain facts which some had hoped to ignore.

First, EPA is needed. The federal government has a legitimate role in the hazardous waste area since the "free market" has not been able to cope with the need for the safe disposal of hazardous wastes. If the federal government's involvement is necessary, then EPA must exist to implement the federal program.

Second, EPA must have adequate resources to do its job. It is ridiculous to expect EPA to meet its current responsibilities and take on new ones if it does not have enough money or trained personnel. The President and Congress must not shortchange the nation's hazardous waste program. The long-term stakes are just too high.

Finally, enforcement is the linchpin of any hazardous waste program. The laws and regulations will be totally ineffective if there is no enforcement. Once again, the critical actor in all this is EPA. Whether the nation's hazardous waste program succeeds or fails will ultimately turn on the adequacy of EPA's enforcement program.

The other contributors have written in more detail concerning the factors behind our present hazardous waste crisis, how the states have addressed this crisis, and how the federal role can be improved. I am confident that the reader will find each article informative and thought provoking.

In closing, I urge each reader to take part in this debate and lend his or her support to the drive to strengthen the nation's hazardous waste program. There can be little doubt in view of the revelations of recent years that we face a major threat to human health and the environment. Action at all levels of government is needed to ensure that the land disposal of hazardous wastes comes to an end and that alternative methods of treatment, recycling, and disposal are utilized to the fullest extent possible.

Congressman James Florio
June 1983

Acknowledgments

The introduction has been adapted from an article, "Beyond Dumping: The Surprising Solution to the Love Canal Problem," which first appeared in the January 1983 issue of the *Washington Monthly*. I thank Phil Keisling for his keen attention to the first drafts of the essay and for his permission to reprint parts of that initial argument.

I am indebted to the Fund for Investigative Journalism and the editors of the *Atlantic Monthly* for their support of the research, travel, and interviews that fed chapter 5. I am especially thankful to Howard Bray of the Fund and C. Michael Curtis of the *Atlantic* for their encouragement when I first proposed researching the higher offices of government influencing the EPA. In financing my research into "The Politics of Toxic Wastes," Bray and Curtis enabled me to interview congressional leaders at the House and the Senate, as well as a number of officials working for the President, during the exciting weeks following Anne Burford's resignation as EPA head.

Chapter 8 has been adapted from an article by the same title, which first appeared in the August 1983 issue of the *Northeast Hazardous Waste Exchange Catalog*. I thank the editors of the catalog for inviting my reflections on European facilities, and I wish them the best in their efforts at recycling wastes.

Parts of chapter 11 first appeared, under the title "Struggling to Be Born," in the spring 1983 issue of the *Amicus Journal*, published by the Natural Resources Defense Council. I thank Peter Borrelli, the editor of *Amicus*, for his continued encouragement of my work and for his thought-provoking assignments.

Avery Comarow, senior editor of *Science 83* magazine, deserves special thanks for commissioning my research into stifled detoxification opportunities. Parts of that research effort, summarized in the afterword, first appeared in the September 1983 issue of *Science 83* magazine and are reprinted by permission of *Science 83* magazine, copyright the American Association for the Advancement of Science. I thank Mr. Comarow for his vigilant skills as an editor and *Science 83* for permission to reprint excerpts from the original report.

I would also like to thank Lynn Taylor, editor of Quorum Books, as well as the rest of the staff at Greenwood Press, for contributing their skills to the making of this book. It has been a pleasure working for this press, and I thank Lynn, especially, who first approached me with this opportunity.

In supplying me with a generous honorarium pool, Quorum Books also enabled me to solicit contributors from a wide range of specialties. This has proven an invaluable aid in the conception of the book. Many thanks also to the chapter contributors, upon whose lived experiences and expertise the foundation of this collection rests. I thank, also, John Rembetski for his review of *Detoxification of Hazardous Waste*, used in the bibliography section, and for our many lively talks while he was a student of mine at Clarkson University.

Finally, I would like once more to acknowledge my debt to Andrea Masters for her constant help and encouragement with this book as with many of my other projects.

Beyond Dumping: An Overview and Introduction

BRUCE PIASECKI

A familiar advertisement by Monsanto proclaims, "without chemicals, life itself would be impossible." Without toxic wastes, life would still be possible, but it certainly would be bleaker.

This isn't an apology for the 77 billion pounds of these wastes that are produced each year in the United States, but a realization that they have become an integral part of our daily life. Sulfuric acid and mercury are inevitable by-products of the pulp and paper industry. The manufacture of life-saving drugs produces zinc and other heavy metals. The textile industry generates toxic dyes and organic chlorine compounds. Even the common doorknob requires electroplating, which generates large volumes of rinse waters and sludges laced with cyanide. For America's major industries, toxic wastes are as common as the garbage trucks that prowl through America's neighborhoods each morning to cart away old newspapers, broken egg shells, and other assorted trash.

This collection of essays, written by expert lawyers, consultants, policymakers, and researchers, is assembled on the following premise: Although Americans generate an unprecedented volume of toxic wastes, we needn't poison ourselves with this stockpile. Each of the following contributors explains key steps we can take to encourage the safe management of our wastes and examines why this step toward the destruction of high-priority wastes should prove profitable to all concerned parties, from corporate managers to citizens legitimately worried about existing standards for toxic waste destruction.

This introductory chapter has been adapted from an article, "Beyond Dumping: The Surprising Solution to the Love Canal Problem," which first appeared in the *Washington Monthly* (January 1983).

Before Americans can shift to these superior means of control, we must understand the factors that sustain the crisis and permit the continued spread of toxic contamination. Part I of this book (chapters 1 to 5) examines these retarding factors. What are the limitations of landfills? Why has it taken so long to learn how to handle toxic wastes? What happens if the Environmental Protection Agency fails to fulfill its congressional mandate? Why has dumping remained the dominant mode of disposal, despite mounting evidence of its threat to public health? Each of these questions is addressed in detail.

For over 40 years, Americans have put their lead, mercury, chlorinated hydrocarbons, PCBs, benzene, cyanides, and other assorted poisons into the ground. For the most part, we have simply dumped them—into landfills, abandoned wells, holding ponds, open fields, and even old Titan missile silos. Then we've crossed our fingers and hoped for the best.

This dump-and-hope approach has prompted people like former California Congressman John Moss to call the problem of toxic wastes "the sleeping giant of the decade" and has led to widespread calls for better dumps. Dumps that are properly identified and policed 24 hours a day—and monitored for generations. Dumps that confine waste in corrosion-resistant containers. Dumps with impermeable liners to prevent groundwater contamination. Dumps that aren't right next door to housing developments, drinking water supplies, and children's playgrounds.

But dumps nevertheless. Yet as toxic waste experts know, the search for the perfect dump is about as fruitful as trying to build a perpetual motion machine. Containers corrode and leak. Rainwater seeps into underground storage areas. Aquifers supplying drinking water eventually become contaminated.

A report by Princeton University's Hazardous Waste Research Program, circulated late in 1982, graphically illustrates the problem: among state-of-the-art landfills that far exceeded current federal standards, some were leaking large amounts of contaminants after only two years of use. Peter Montague, a researcher at Princeton University's Center for Energy and Environmental Studies, was a principal author of this effort. In chapter 1, Montague summarizes the results of recent landfill research and analyzes its implications.

Montague begins by describing engineering efforts to upgrade landfills, so as to secure them for toxic wastes. Next, he explains why even these best dumps leak. In its never-ending uphill battle against leaking landfills, the EPA has proposed various remedial strategies. The central sections of Montague's essay present new evidence on why these legislated requirements for monitoring and treating contaminated groundwater may prove ineffectual. Faced with the fact that our best landfills are as leakproof as colanders, Montague concludes by itemizing a series of corrective actions that must be taken.

The lesson in all this is obvious but usually overlooked. The last thing we should do with our toxic wastes is figure out the best way to dump them. Instead, we should do something safer and far more sensible: turn them into harmless chemicals and make some money selling what's left over.

But learning how to do this, and then regulating industry to make the shift from dumping, is not easy. In chapter 2, Gary N. Dietrich reviews the learning curve hazardous waste experts at the EPA experienced over the last decade, and explains the difficulties they encountered in their efforts to arrest growing problems. In a field where most experts have less than five years' experience, Dietrich should know. A thorough, exacting Cal Tech engineer, Dietrich was with the EPA from its birth in 1970 until 1982. He organized the first Office of Toxic Substances; solicited the first survey that led to the creation of the Superfund, a national account for the emergency clean-up of dumps; and for several years—during the period of peak growth at the EPA—he managed the EPA's entire budgeting process. Now president of Clement Associates, a science and engineering consulting firm, Dietrich argues that Americans are still only at the midpoint in the learning curve. This lack of reliable information plays a major role in sustaining the crisis, according to Dietrich.

Yet to others, it is politics—not technical know-how—that prevents Americans from moving beyond dumping. In chapter 3, Khristine L. Hall, a senior attorney for the Environmental Defense Fund, argues that the EPA has sufficient information to act but is often delayed by political considerations. For Hall, this is all the more reason why private watchdog agencies must monitor the EPA to keep it vigilant. By reconstructing an insider's account of the legal actions various private groups took against the EPA, Hall examines two recent sources of controversy—efforts to improve dumps by stipulating standards for construction and efforts to limit leachate by restricting the dumping of toxic liquids into landfills.

Chapter 4 also tells us much about the making of toxic waste regulations in America. Written by James Banks, a senior attorney for the Natural Resources Defense Council, this chapter examines efforts to control toxic discharges into American waterways. Banks patiently describes the evolution of America's policies across 14 sessions of Congress, five administrations, and extensive litigation in the federal courts. As a result, the reader is left with a clearer understanding of the means by which rule making occurs in America. One watches the ways in which policymakers, lawyers, technicians, and representatives from the regulated industries walk to the middle of the tightrope and at last shake hands in agreement. Yet Banks' chapter contains evidence of a larger complicating factor: even the hard-earned results of due process can be cut, at times, by the whim of a new administration. Banks' essay is a telling test case on the ways in which toxic waste legislation has become intimately entangled with presidential efforts at regulatory relief.

Chapter 5 joins the concerns of the preceding chapters into a single narrative. By examining the process by which EPA regulations are checked and revised by the Office of Management and Budget (OMB), it discloses the role of federal politics in toxic waste controls. In this essay, efforts to upgrade dumps, control toxic effluents from steel and battery manufacturers, and design cost-effective regulations are used to assess a new kind of rule making now ascending in Washington. The fate of America's toxic waste programs is intimately linked with these reviews of EPA regulations by the OMB. The editor examines two complex episodes in hazardous waste rulemaking by the EPA that demonstrate the ways in which the EPA administrator still is controlled by OMB regulatory managers working for the President.

Despite these factors sustaining the crisis, detoxification is possible and efforts to shift beyond dumping have increased in tempo.

Before dismissing this as a modern day version of the alchemist's lead-into-gold fantasy, consider one thing: detoxification of hazardous waste happens to be standard operating procedure in much of the rest of the industrial world. In the Netherlands, the Chemical Waste Act of 1976 explicitly prohibited the dumping of a wide range of toxic wastes, a ban that has since spawned a thriving waste detoxification business. In Denmark, the Kommunekemi plant handles every toxic waste in the country. And if it can't destroy or recycle a particle poison, the Danes don't dump it—they put it in storage until they can find a way to treat it.

Holland and Denmark are not exactly world-class industrial powers—like West Germany, for example. But those who doubt that detoxification is both effective and economical need only visit any of that country's 15 waste treatment facilities, which are operated as part of a coordinated national program. In the West German state of Bavaria, for instance, which is about the size of Ohio, 85 percent of its toxic wastes are destroyed rather than dumped. To them, detoxification has virtues beyond merely protecting their citizens: the West Germans recover heat and materials and generate electricity from their facilities. Chapter 8 provides further information on European facilities.

These European examples haven't gone unnoticed in some parts of the United States. In 1981 California Governor Jerry Brown appointed a special Toxic Waste Group to examine alternatives for the 1.3 million tons of hazardous wastes disposed annually in the state's landfills. Prepared in cooperation with representatives from Dow Chemical, Friends of the Earth, and various academic institutions, the report concluded that 75 percent of California's toxic wastes could economically be recycled, treated, or detoxified. California since has established a low-interest loan program for building detoxification facilities; to discourage dumping, the state has instituted new dumping rules and increased the landfill tax by 600 percent. Hit with additional charges of $4 million per year, Chevron,

Getty, and similar companies are now dumping far less waste in California.

Before you assume California's anti-dumping crusade will cause many beleaguered manufacturers to throw in the towel, consider chapter 6. Written by Gary A. Davis, a principal author of California's study, this chapter summarizes the origins, development, and implementation of California's program. This program can serve as a valuable model for many states now struggling with local resistance to dumping. In demonstrating that the knowledge is available, the tools have been built, and the shift from land disposal depends upon political will, Davis' work indicates that there are ways by which the entire nation can rectify its toxic predicament by the end of the decade.

Meanwhile, most state programs don't come close to the California example. In chapter 7, attorney Sheila Brown explains why. After producing three times as much toxic waste per citizen as West Germany, American industry continues to dump 80 percent of its wastes into pits, ponds, surface impoundments, or landfills. Much of these wastes work their way toward American water supplies. Less than 20 percent is treated in any way, and the most common treatment method—incineration—is often conducted under uncontrolled conditions that transfer the poisons into the atmosphere (see afterword).

For a country that legitimately prides itself on its ability to mobilize rapid deployment forces and sophisticated space programs, this continued enthusiasm for dumping is not only appalling, but downright mystifying. This is especially so when one considers the spread of toxic contamination into our groundwater supplies.

Assembling evidence from the first inventories by EPA's regional offices, Brown reveals that every state in the nation has abandoned at least one regional drinking water supply because of toxic contamination. The link is direct—the contamination originates from leaking landfills or inexpert dumping. Next, Brown argues that this problem, clearly of national dimensions, deserves a coordinated federal control effort. State controls, the record indicates, are simply not sufficient.

It has been five years since Love Canal was declared a national emergency, yet Americans still lack an effective toxic waste treatment market. Americans now generate over three times as many poisonous wastes as existing permitted facilities could treat if functioning at full capacity. New York State alone annually generates over 400,000 tons of toxic wastes beyond its present treatment capacity.

The story behind America's inability to create a reliable treatment market is a classic tale of how regulatory neglect allows market failure. The solution to the American crisis does not await the glamour of future technologies, or a revolutionary halt to our synthetics production. The flaw has been in our thinking and planning, not in our machines.

The final section consists of three essays that propose remedies for these managerial flaws. Written by a prominent economist, a federal researcher, and the editor, these chapters signify a series of important developments since Love Canal. A long-standing logjam between the industries handling wastes and the policymakers regulating the industries has begun to loosen. Herein lie perhaps the most reliable grounds for hope. For when economists, big business, and federal officials speak openly and share information effectively, American citizens can expect a more workable climate for charting the path toward the safer management of wastes.

In chapter 9, Patrick G. McCann, an economist for the consulting firm Booz-Allen & Hamilton, describes the ways he advises leading corporate managers about the inevitability of stronger controls. By describing trends in the regulatory climate against dumping, McCann emphasizes the importance of anticipation and preparedness. According to McCann, the next five years will result in a new wave of regulations which demand that managers evaluate a range of critical non-economic factors in waste disposal.

Serving as a primer for those interested in the political economy of toxic controls, chapter 9 includes such factors as liability protection, technical reliability, public service, and public image in its decision models. In order to keep on top of the situation, a professional manager can no longer simply resort to the cheapest dumping options. McCann concludes by providing a prospectus of new regulatory trends and by itemizing the most useful business response to these trends.

Understanding the fierce logic of policymaking, Joel S. Hirschhorn writes in chapter 10 about the role of market incentives for improving the management of wastes. As the director of a three-year study by the U.S. Congress' Office of Technology Assessment (OTA), Hirschhorn led Washington's research efforts to determine the most effective ways to equalize the costs between cheap but hazardous dumping and safe but more costly treatment techniques. The results of this work are summarized in chapter 10. (For those interested in the full report, see *Technologies and Management Strategies for Hazardous Waste Controls*, published in March 1983 by OTA; also see appendix A for an extensive excerpt.)

Ironically, the most forceful evidence for stronger toxic waste regulations has come most recently from the corporate sector. Companies that have tried to capture the treatment market but continue to be hurt by regulatory loopholes that permit their competitors to mismanage wastes at far less expense are now fighting back. Chapter 11 reports on these new industrial efforts to stabilize America's treatment market. Written by the editor from extensive interviews with members of the Hazardous

Waste Treatment Council (HWTC), this essay represents the first public account of industry's strong interests in shifting America beyond dumping.

Early in 1982, a significant number of American waste handlers assembled in Washington to form a lobbying council. Fearing federal backsliding in regulating wastes, HWTC incorporated the same month and began filing suits against the police for failing to arrest known criminals within their own industry: for every time a dumper opened his spigot illegally, those trying to treat wastes lost a part of their market.

This industrial alliance holds much promise for the future, for it gives economic clout to concerns about public health. As HWTC's client list continues to grow, the council already represents ENSCO, SCA Services and CECOS—three of the top seven American firms handling 70 percent of the toxic waste treatment market. The editor describes the ambitious plans these companies have to turn a profit by freeing the public of its waste stockpiles. A fuller account can be found in appendix C, where the current executive director of HWTC explains both the most current positions articulated by HWTC in Washington and the ways in which innovative industries present new solutions to our 30-year-old waste problems.

Clearing the air of the suspicions that surround the toxic waste crisis will take years, yet the evidence presented in these chapters shows that rapid progress can be made. In the afterword, the editor reflects on these new grounds for hope by reporting on a few of the hundreds of technical breakthroughs which make a shift from dumping not only preferable but also profitable once reliable regulations are established. For without the reforms indicated in these essays, America's impressive detoxification arsenal will remain idle—tokens of practical solutions lost in our labs because of a blind trust in dumping.

Part I
FACTORS SUSTAINING THE CRISIS

1
The Limitations of Landfilling

PETER MONTAGUE

In the United States, approximately 80 percent of all hazardous wastes end up in landfills [#32, p. 250]. Although there are other ways now available for managing dangerous industrial byproducts—such as waste reduction, recycling, high-temperature incineration, and detoxification [#70; #57]—the bulk of America's toxic wastes still are buried in a hole in the ground. There are two key reasons for this: (1) from an engineering viewpoint, it is relatively simple to do; and (2) federal regulations make it legal to pass the major costs of landfilling on to the general public. As a result, landfilling is by far the cheapest waste disposal alternative and is very hard to resist.

THE DESIGN OF A MODERN LANDFILL

The practice of landfilling is as old as humanity itself. For more than a million years humans have thrown their trash onto the earth, thus creating midden heaps, or dumps. For most of history, this pratice has sufficed. During the present century, however, nations began dumping materials that are increasingly hazardous and that do not easily break down in the environment. Thus, our old habit of throwing our trash in (or onto) the earth has begun to cause a new set of problems, chief among them being contamination of our underground drinking water supplies [#10]. Over the last three or four decades, civil engineers have improved landfill siting and design to minimize these problems. The modern landfill has become, as a result, a carefully engineered facility aimed at minimizing any hydraulic connection between wastes and the environment [#50; #15; #19; #58].

A modern hazardous waste landfill has four essential components: (1)

an appropriate or inappropriate natural geologic setting, (2) one or more liners below the wastes, (3) a final cover over the wastes, and (4) a leachate collection system. The natural geologic setting can be selected to minimize the possibility of wastes contacting groundwater. The other three features must be engineered. The liner, or liners, below the wastes may be made of clay or of some synthetic membrane (such as a sheet of plastic film). The liner essentially creates a bathtub in the ground. There are two ways wastes can get out of the bathtub—by leaking out the bottom through a hole or by the bathtub filling up and spilling over its sides.

To prevent the bathtub from filling up with rainwater, a final cover is placed over the wastes when the landfill is full. The final cover is generally constructed of the same material as the liner beneath the wastes. This final cover acts as an umbrella, keeping rain and snow out of the bathtub. The leachate collection system is a set of pipes, placed in the lowest location in the landfill, which acts as a drain. If liquids enter the landfill (or are created or released by chemical interactions within the landfill), the liquids drain into the leachate collection system; from there they are removed (usually by a pump) and usually sent to a wastewater treatment facility.

The leachate collection system only operates so long as the pumps function to remove leachate and the pipes remain unclogged. When the leachate collection system fails, any liquids in the landfill will begin to fill the bathtub, eventually overflowing its sides.

WHY EVEN SECURE LANDFILLS LEAK

During the 1970s, it became clear that conventional landfills leak [#48; #53; #54; #67] and that landfill leakage can cause serious contamination of soil and water. As a result, American engineers set about designing new landfills. These quickly became known as "scientific" landfills or "secure" landfills. Both government officials and private consulting engineers developed a religious faith in these new technologies. For example, Thornton W. Field, a special assistant to the administrator of the U.S. Environmental Protection Agency (EPA) said, in early 1982, that the EPA's latest regulations for chemical landfills would "adequately protect human health and the environment . . . basically forever" [#47, p. 1]. In mid-1983 a large midwestern engineering firm asserted that its landfill design would "completely" and "permanently" isolate chemical wastes from the environment [#69, pp. 4, 9, 14].

Such sweeping claims are not isolated aberrations, but reflect a deep and abiding faith among many engineers and regulatory officials that "secure" landfills can permanently contain chemical wastes. The Chemical Manufacturers Association has nurtured this view with a series of full-page advertisements in magazines such as the *New Yorker* [#13, p. 145; #14, p. 179].

Many citizens and engineers are now asking: Is it really possible to design a leakproof landfill? The answer to this question depends upon the answer to another key question: For how long into the future do we want the landfill to be leakproof? If we are willing to expend enough money and brains and materials on design and construction and maintenance, there is a good chance that we can create a landfill that will maintain its integrity (that is, won't leak) for 10 to 40 years. But if we want our landfill to maintain its integrity into the indefinite future, the answer is certainly no, we cannot do it.

How can we be so certain about this answer? How do we know that all landfills must eventually leak? We know this with absolute certainty because of one of the most fundamental laws of the universe, understood by chemists and physicists as the second law of thermodynamics.

The second law began to be understood in the middle of the nineteenth century, when it was first applied to heat engines [#3; #36, pp. 34–36]. The simplest and most succinct statement of the second law is this: disorder in a system and its surroundings always increases. In this context, the proper name for disorder (or at least for measuring disorder) is entropy; the second law says that the entropy of a system and its surroundings never spontaneously decreases. Like time itself, entropy only flows in one direction: it increases. This is a fundamental and immutable law of the universe.

Although it is a highly abstract concept, the second law of thermodynamics has very important (and inescapable) implications for toxic waste management. When you drop a teacup onto an unyielding floor, the cup spontaneously smashes into many pieces. You never see the reverse happen—you never see a smashed teacup spontaneously rearrange itelf into a whole teacup again. You never see a person getting younger instead of older. You never see the clothing in your bedroom or the books and papers in your office spontaneously rearrange themselves into a more orderly grouping than you last left them in. Stated another way, the second law says that, left to itself, a system becomes as disordered as possible [#5].

In a localized area, humans can temporarily reverse the spontaneous direction of the universe by applying energy to a particular situation. We can bring order out of disorder by exerting ourselves. (Life itself is a temporary reversal of the natural tendency of disorder to increase; however, as we all know, the highly ordered arrangement of molecules that constitutes a living organism is only temporary—eventually our highly ordered systems revert to dust and tend to become distributed throughout the environment.) Even in these cases where we *locally* reverse the flow toward disorder, our exertions always, themselves, increase the disorder in some other part of the universe—and so the second law is satisfied: taking into account the system (the local area where order is increasing) plus all of the area's surroundings, entropy is always increasing, never decreasing. This law explains what will happen to chemicals placed in a landfill. They will tend—in the long run—not to stay where we put them. They will tend to become dispersed throughout the environment. They will tend to spread, to move into the atmosphere, and into the soil, and into the groundwater.

When will this dispersal happen? It is impossible to say exactly when it will happen. But we can say with certainty that it will *begin* to happen as soon as humans stop managing the site, stop expending energy to counter the tendencies of the second law. As soon as the site is left to nature, natural forces will take over, the site integrity will begin to deteriorate, and the buried chemicals will begin to migrate. As we saw earlier, a secure landfill has two essential parts—a bathtub and an umbrella. Both of these essential components are subject to the second law.

The bathtub (the liners at the bottom of the landfill) will tend to deteriorate as time passes. If they are synthetic membranes, their molecules will eventually separate from each other and go their different ways. Weak spots will develop, cracks will develop, holes, tears. All these materials have a finite lifetime and they will eventually degrade. During the long process of setting standards for landfills, the federal EPA acknowledged this fact:

> Some have argued that liners are devices that provide a perpetual seal against any migration from a waste management unit. EPA has concluded that the more reasonable assumption, based on what is known about the pressures placed on liners over time, is that any liner will begin to leak eventually. [#63, pp. 32284–85]

After the liner deteriorates, gravity will tend to pull the wastes downward, toward the groundwater. Diffusion will tend to disperse the wastes spontaneously in many directions. The ordered arrangement of wastes will become increasingly disorderly and dispersed.

The umbrella is more subject to natural deterioration than the bathtub. This is because the umbrella is up on the surface of the land where it is subject to relentless natural agents:

1. Rain, snow, hail, sleet, and wind will erode the umbrella.
2. The freeze-thaw cycle continually works to loosen and degrade the umbrella.
3. When these agents expose the umbrella to the sun, its clay will begin to dry out and crack, or its synthetic liner will deteriorate from heat and ultraviolet radiation.
4. Trees and other vegetation that send down taproots will penetrate the umbrella.

5. Rabbits, woodchucks, mice, moles, voles, ants, snakes, and other soil-dwelling creatures will tend to burrow into the umbrella. Eventually they will penetrate it, weakening or destroying it.
6. Humans in their marvellous unpredictability will sooner or later forget where the umbrella was located, and sometime after that they will build trailer parks, roads, homes, schools, factories, or other structures on the site. They will flatten the site with a bulldozer or cut into it with a backhoe. When these things happen, the wastes will begin to disperse throughout the environment, driven by inescapable forces governed by the second law of thermodynamics.

It is not possible to say which combination of these six elements will ultimately have the greatest effect in destroying the cap of a landfill. It *is* possible to say, however, with a very high degree of confidence, that sooner or later the cap of any landfill will be destroyed. When that happens, the bathtub will fill with rainwater, spill over its sides, and contaminate the surrounding environment.

How long will the chemicals in a landfill remain dangerous? Answer: for a very long time; in some cases, for decades; in other cases, for hundreds of years; in some cases, basically forever. Even chemicals that can degrade will do so very slowly in a landfill environment where the earth is cool and oxygen is in short supply. And, of course, many common contaminants—like arsenic, mercury, beryllium, lead, chromium and cadmium—will not degrade at all.

Since it is impossible to accurately predict future patterns of land use, water use, and habitation by humans, prudent planning requires that we assume the worst: whenever we dump toxic chemicals, someone, somewhere, will be harmed.

WHY AMERICA'S NEW REGULATIONS ARE INSUFFICIENT

The federal Environmental Protection Agency spent six years issuing landfill regulations. These regulations, which became effective January 1983, recognize three periods in the lifetime of a landfill: the operational period while the landfill is being filled; the closure period when the cap (umbrella) is being installed; and the post–closure period when nothing much is going on except routine maintenance and monitoring to see when the landfill begins to leak.

The length of the operational period for a landfill is determined by commercial factors: What is the capacity of the landfill to begin with, and how fast can it accept waste? Most landfills vary in size from 5 to 500 acres and operate for 10 to 20 years. The length of the closure period is established by engineering considerations: How fast can the cap (umbrella) be constructed? (Two months to two years is a typical closure period.) The length of the post–closure period is established by law. Currently, it is 30 years.

At the end of the post–closure period, the owner/operator of the landfill is officially absolved of responsibility; after 30 years, maintenance and monitoring becomes the responsibility of the federal government (and thus the burden of all taxpayers). There is one possible exception to this situation: if a landfill starts leaking before the end of the 30-year post–closure period, the regional administrator of the EPA can extend the length of the post–closure period for an indefinite time. Thus, the law contains inducements that will favor construction of landfills that will not leak during the operation, closure and 30-year post–closure periods, but that will begin to leak soon thereafter. In response to this predicament, the current regulatory scheme has two major parts: a liquids management strategy to put off that time when leakage occurs and a remedial strategy to clean up the leakage after it occurs.

The liquids management strategy assumes that, after the closure period, it is the cap (umbrella) that becomes the chief protective component of the strategy [#63, p. 32285]. The cap is intended to prevent fluids from entering the landfill from the top, thus minimizing the formation of leachate during the post–closure period.

Why Remedial Strategies May Prove Ineffective

In addition to a liner and a leachate collection system, new landfills must be constructed with groundwater monitoring wells located at the edges of the landfill. Groundwater is a body of water found throughout the earth at an average depth of about 30 feet below the surface of the land [#21]. The topmost surface of a body of groundwater is called the "water table." In some places (in arid and semi-arid regions, such as the southwestern U.S.) the water table may be several hundred feet below the surface of the land. In other places, the water table may be right at the surface (we call such a place a swamp). In the wet regions of the country (any place east of central Kansas), the water table is typically only 2 to 30 feet below the surface of the land.

Since half the drinking water in the United States is pumped from groundwater reserves, the contamination of groundwater by chemicals is a serious and difficult problem [#10]. Current federal landfill regulations require that a minimum of three monitoring wells be placed around any new landfill. Samples must be drawn from these wells periodically and chemical tests conducted to see whether the landfill has begun contaminating groundwater.

Federal regulations assume that, in the same way that the liner will eventually fail, so too the cap will eventually fail: "Caps, like bottom liners, cannot be expected to last forever" [#63, p. 32314]. If failure occurs before the end of the 30-year post–closure period, the owner/operator is then required to institute a remedial pumping program. This consists of drilling wells down into the groundwater and pumping the water to the surface, where it is treated to remove any contamination and then returned to the ground. If remedial action begins before the end of the post–closure period, the owner/operator must continue the program until he or she can demonstrate that, for three consecutive years, the water being pumped to the surface is as clean as the area's groundwater was before the landfill went into operation. (If, however, the need for remedial action appears *after* the end of the post–closure period, when the federal government has taken responsibility for the site, the law is silent on how long such a remedial pumping program must be continued, or whether, in fact, one would be required at all.)

This "remedial strategy" assumes that pumping groundwater will make it possible to remove contaminants and cleanse the water. In some cases this may be true but in others it will almost certainly not be true. The earth is a very complicated place. Down in the regions where groundwater typically flows, the rocks and soil may not be simple or easy to understand. The rate and direction of groundwater flow may change from season to season, affected by conditions at the surface. In a few sites, groundwater may flow through pure sand. In these cases, groundwater flows may be easy to understand and predict. Then the proper placement of groundwater monitoring wells, and remedial pumping wells, will be straightforward. In a fractured geologic medium, however, one could only predict groundwater flow if one had a full understanding of the pattern of fractures—how big they were and where they went. Without a detailed understanding of the pattern of fractures, one would not know where to place monitoring wells or where to place remedial pumping wells. The fractures can be thought of as pipes with water flowing through them. In order to drill a well (which is typically 2″ to 4″ in diameter) and intersect one of the pipes, one would have to have a very good idea of where the pipe was. Such detailed knowledge is impossible to acquire without drilling many wells into the rocky ground—a very expensive hit-or-miss proposition.

Corrective Options

It is apparent, then, that the federal government's regulatory program is suspect. First of all, it is premised on the certainty that, sooner or later, all landfills will leak. But it is further premised on the highly uncertain proposition that all the groundwater contaminated by leaking landfills can be retrieved and cleaned up at reasonable cost. As we've seen, groundwater flowing through fractured geologic media very likely cannot be cleaned up at any cost. And even when the geologic medium is fairly simple (as in the case of sand), the costs of pumping millions of gallons of water year after year after year may exhaust the financial resources available to do the job. David Miller of the hydrological consulting firm of Geraghty & Miller has estimated that the average cost of partially cleaning up a contaminated groundwater site will run between $5 and $10 million. Mr. Miller further estimates that the total restoration of a badly contaminated aquifer (underground water source) to potable quality could take "decades" and might cost $500 million to $1 billion or more [#22].

Considering these difficulties, what can be done?

There are two basic approaches that can be taken: (1) improve landfill siting and design to minimize the expected problems or (2) restrict the kinds of materials that can be buried in the ground.

Current federal regulations have taken the first approach. But they have not gone as far as they might. One could, for example, prohibit the siting of landfills over fractured geologic media. David Miller has recommended such a prohibition in testimony before Congress [#36]. One could also require that a landfill site be "simple"—simple enough to be modeled mathematically with high confidence, so that one could then predict the directions and rates of groundwater flow.

One could go further and require that no part of a landfill be constructed below ground. An "aboveground" landfill would be like a football stadium built on the surface of the ground. The wastes would be placed on the "playing field" and then, as the facility filled up, the wastes would be put up in the bleachers. The important advantage of such a landfill would be (1) its leachate collection system could operate by gravity alone without any need for pumps; (2) such a facility would be relatively easy to monitor for leakage; and (3) its very presence would serve as a continual reminder of its existence and the hazardous threat it contains. Kirk Brown of Texas A & M University, one of the country's leading researchers on landfill problems, has suggested the use of aboveground facilities, as have some researchers in the waste management industry [#8; #7; #6].

Finally, one could require a twin liner system. One liner could be made of synthetic membrane and one liner could be made of 5 to 10 feet (or more) of clay. By this means, one might further reduce the rate at which leakage occurred.

None of these measures will change the ultimate fact that landfills leak. However, these measures might reduce the rate at which leakage occurs, or might make it easier to detect leakage or to clean up leakage that had been detected.

To *prevent* the leakage of hazardous chemicals out of landfills, we will have to stop putting hazardous chemicals into landfills. This does not mean that we must end all landfilling. Far from it. It only means that we must end the landfilling of *hazardous* materials.

This is the approach being taken in California, where the state Department of Health Services has been formulating regulations since October 1981, when Governor Brown ordered the Department to "prohibit the land disposal of highly toxic wastes." (See chapter 6 of this book for more on the California example.)

Illinois is taking yet another approach to the problem of banning hazardous wastes from landfills. By Public Act 82–572, Section 39 of the "Environmental Protection Act of 1970" has been amended to read as follows:

> Commencing January 1, 1987, a hazardous waste stream may not be deposited in a permitted hazardous waste site unless specific authorization is obtained from the [Illinois Environmental Protection] Agency by the generator and the disposal site owner and operator for the deposit of that specific hazardous waste stream. The Agency may grant specific authorization for disposal of hazardous waste streams only after the generator has reasonably demonstrated that, considering technological feasibility and economic reasonableness, the hazardous waste cannot be reasonably recycled for reuse, incinerated or chemically, physically, or biologically treated so as to neutralize the hazardous waste and render it nonhazardous.

This decision sets a precedent worthy of comprehensive emulation.

There is no such thing as a secure landfill; there is no such thing as "permanent" disposal of hazardous wastes in the ground. There is "long-term storage" in the ground but sooner or later what we put in the ground will move—and the place it is most likely to move to is someone's drinking water supply.

Nevertheless, many federal officials sustain a religious faith that landfilling can solve our disposal problems. In recent years, federal regulations have seesawed between insistence that clay and synthetic membrane liners can protect us and the explicit recognition that ultimately leakage will occur [#60; #59; #61; #62; #63; #64; #65; #66]. There is now, however, sufficient theoretical support from science and empirical data from laboratory and field studies to force the conclusion that both landfill liners and caps will ultimately leak, whether they are made of clay [#2; #1; #4; #16; #25; #26; #33; #34; #38; #39; #40; #41] or of synthetic membranes such as rubber or plastic [#28; #29; #38; #39].

The costs of continued landfilling are high, both in terms of dollars and in terms of health costs. The $1.6 billion Superfund—a federal program to clean up abandoned dumps—is just the tip of the iceberg. In New Jersey alone, the 300-odd abandoned dumps could cost $1.6 billion to clean up (depending on what level of "clean" we want cleanup to achieve). Even when we are willing to live with considerable residual contamination, cleanup of an old hazardous waste landfill typically costs tens of millions of dollars.

Furthermore, it is our continued reliance on landfilling that prevents the use of already existing alternatives. This fact is recognized by government officials [#9; #31; #57] and by private businessmen [#17; #43].

So long as landfilling is available as the cheap and easy option, who will use more complex (though surer and safer) disposal options?

The federal government has taken no significant initiatives to discourage landfilling, but state governments are beginning to act. These state actions, combined with aggressive lawsuits by individuals, to collect damages from landfill owners, operators, and users will in time spell the end of dumping toxic wastes in America. Until then, new landfill proposals, and proposals to expand old landfills, will—and should—meet with vigorous public opposition. Where governments will not protect the people, the people will have to take steps to protect themselves.

REFERENCES

1. Anderson, David, and K. W. Brown. "Organic Leachate Effects on the Permeability of Clay Liners." In *Land Disposal: Hazardous Waste; Proceedings of the Seventh Annual Research Symposium* [EPA–600/9–81–002b]. Washington, DC: U.S. Government Printing Office, 1981, pp. 119–30.
2. Anderson, David; K. W. Brown; and Jan Green. "Effect of Organic Fluids on the Permeability of Clay Soil Liners." In *Land Disposal of Hazardous Waste; Proceedings of the Eighth Annual Research Symposium* [EPA–600/9–82–002]. Washington, DC: U.S. Government Printing Office, 1982, pp. 179–90.
3. Angrist, Stanley W., and Loren G. Hepler. *Order and Chaos; Laws of Energy and Entropy*. New York: Basic Books, 1967.
4. Barrier, R. *Zeolites and Clay Minerals and Sorbents and Molecular Sieves*. New York: Academic Press, 1978.
5. Bent, Henry A. "Haste Makes Waste; Pollution and Entropy." *Chemistry*, vol. 44 (October 1971), pp. 6–15.
6. Bridge, Dan; Larry Graybill; Ed Hillier; Al Van Tassel; Ned Winders; and Douglas Carter. "On-Site Remedial Strategies for Economically Non-disposable, Non-treatable Hazardous Wastes, with Emphasis on Aboveground Cells." Unpublished typescript available from Rollins Environmental Services, Inc., P.O. Box 609, Deer Park, TX 77536.
7. Brown, K. W. "Landfills of the Future." Unpublished; December 15, 1982. Available in typescript from the author at Soil and Crop Sciences Department, Texas A & M University, College Station, TX 77843.
8. Brown, Kirk W. "Testimony before the [U.S.] House [of Representatives] Subcommittee on Natural Resources, Agriculture Research and Environment of the Committee on Science and Technology, November 30, 1982." Unpublished. Available in typescript from the subcommittee.
9. Bulanowski, George A.; Greg H. Lazarus; Larry Morandi; and Jonathan M. Steeler. *A Survey and Analysis of State Policy Options to Encourage Alternatives to Land Disposal of Hazardous Waste*. Denver, CO: National Conference of State Legislatures, 1981.
10. Burmaster, David E., and Robert H. Harris. "Groundwater Contamination: An Emerging Threat." *Technology Review* (July 1982), pp. 51–62.

11. [California] Department of Health Services. "Discussion Paper: State Action to Reduce Land Disposal of Toxic Wastes." Sacramento, CA: Department of Health Services, January 1982.
12. [California] Department of Health Services. "Follow-up Discussion Paper: State Action to Reduce Land Disposal of Toxic Wastes." Sacramento, CA: Department of Health Services, May 1982.
13. Chemical Manufacturers Association. "Managing Chemical Wastes." *New Yorker* (December 15, 1980), p. 145.
14. Chemical Manufacturers Association. " 'My Job Is Managing Chemical Industry Wastes. What I Do Helps Make the Environment Safer Today—and for Generations to Come.' " *New Yorker* (September 14, 1981), p. 179.
15. Cheremisinoff, Nicholas P.; Paul N. Cheremisinoff; Fred Ellerbusch; and Angelo J. Perna. *Industrial and Hazardous Wastes Impoundment.* Ann Arbor, MI: Ann Arbor Science, 1979.
16. Daniel, David E. "Predicting Hydraulic Conductivity of Clay Liners." Unpublished. Available in typescript from the author at Department of Civil Engineering, University of Texas, Austin, TX 78712.
17. Durning, Marvin B. "Statement of Marvin B. Durning Representing the Hazardous Waste Treatment Council before Subcommittee on Commerce, Transportation and Tourism, of the Committee on Energy and Commerce, U.S. House of Representatives, March 31, 1982. In U.S. House of Representatives, *Resource Conservation and Recovery Act Authorizations for Fiscal Year 1983 and Fiscal Year 1984* [97–169]. Washington, DC: U.S. Government Printing Office, 1982, pp. 272–85.
18. Farb, Donald G. *Upgrading Hazardous Waste Disposal Sites* [SW–677]. Washington, DC: U.S. Government Printing Office, 1978.
19. Fields, Timothy, Jr., and Alfred W. Lindsey. *Landfill Disposal of Hazardous Wastes: A Review of Literature and Known Approaches* [SW–165]. Washington, DC: U.S. Government Printing Office, 1975.
20. Fischback, Bryant. "Statement of Dow Chemical U.S.A. Presented to the [California] Department of Health Services Workshop February 19, 1982, and the Subcommittee on Commerce, Transportation and Tourism, Committee on Energy and Commerce, U.S. House of Representatives, March 31, 1982." Available from the author at Dow Chemical, P.O. Box 1398, Pittsburgh, CA 94565.
21. Garrels, Robert M.; Fred T. Mackenzie; and Cynthia Hunt. *Chemical Cycles and the Global Environment; Assessing Human Influences.* Los Altos, CA: William Kaufmann, 1975.
22. Geraghty & Miller, Inc. "Cleanup of Contaminated Groundwater Costs $5 to $10 Million Per Site." Syosset, NY: Geraghty & Miller, Inc., July 1982.
23. Geswein, Allen J. *Liners for Land Disposal Sites* [EPA/530/SW–137]. Washington, DC: U.S. Government Printing Office, 1975.
24. Green, William J.; G. Fred Lee; and R. Anne Jones. "Clay-Soils Permeability and Hazardous Waste Storage." *Journal of the Water Pollution Control Federation*, vol. 53 (1981), pp. 1347–54.
25. Green, William J.; G. Fred Lee; and R. Anne Jones. "Final Report [to the U.S. Environmental Protection Agency] on Impact of Organic Solvents on the Integrity of Clay Liners for Industrial Waste Disposal Pits: Implica-

tions for Groundwater Contamination." Fort Collins: Colorado State University Department of Civil Engineering, 1979.

26. Green, William J.; G. Fred Lee; R. Anne Jones; and Ted Palit. "Interaction of Clay Soils with Water and Organic Solvents: Implications for the Disposal of Hazardous Wastes." *Environmental Science & Technology*, vol. 17 (May 1983), pp. 278–82.
27. Grim, R., and F. Cuthbert. *The Bonding Action of Clays*. Report of Investigation no. 102. Champaign: Illinois State Geologic Survey, 1945.
28. Haxo, H.E., Jr. "Durability of Liner Materials for Hazardous Waste Disposal Facilities." In *Land Disposal: Hazardous Waste; Proceedings of the Seventh Annual Research Symposium* [EPA–600/9–81–002b]. Washington, DC: U.S. Government Printing Office, 1981, pp. 140–56.
29. Haxo, H. E., Jr. "Effects on Liner Materials of Long-term Exposure in Waste Environments." In *Land Disposal of Hazardous Waste; Proceedings of the Eighth Annual Research Symposium* [EPA–600/9–82–002]. Washington, DC: U.S. Government Printing Office, 1982, pp. 191–211.
30. Hirschhorn, Joel M. "Statement [November 30, 1982] of Joel M. Hirschhorn, Project Director, Office of Technology Assessment, before the Subcommittee on Natural Resources, Agriculture Research and Environment, Committee on Science and Technology, United States House of Representatives." Unpublished. Available in typescript from the subcommittee.
31. Hirschhorn, Joel M., and others. *Technologies and Management Strategies for Hazardous Waste Control* [OTA–M–196]. Washington, DC: U.S. Government Printing Office, March 1983. [Also, see chapter 10 and appendix A of this text.]
32. Josephson, Julian. "Hazardous Waste Landfills." *Environmental Science & Technology*, vol. 15 (March 1981), pp. 250–53.
33. Kingsbury, Robert P. "Position Paper; Chemical Resistance of Clays." 1982. Unpublished typescript available from the author at American Colloid Co., P.O. Box 696, Laconia, NH 03246.
34. Lee, G. Fred, and R. Anne Jones. "Draft; Hazardous Waste Disposal by the Clay Vault Method: Is It Safe?" Occasional Paper no. 79, Department of Civil Engineering, Environmental Engineering Program. Fort Collins: Colorado State University, 1982.
35. Lutton, Richard J. "Selection of Cover for Solid Waste." In *Land Disposal of Hazardous Wastes, Proceedings of the Fourth Annual Research Symposium* [EPA–600/9–78–016]. Washington, DC: U.S. Government Printing Office, 1979, pp. 319–25.
36. Miller, David W. "Testimony by David W. Miller before the Subcommittee on Natural Resources, Agriculture Research and Environment [of the Committee on Science and Technology, U.S. House of Representatives] Hearing November 30, 1982." Unpublished. Available in typescript from the subcommittee.
37. Miller, G. Tyler. *Living in the Environment*, 2d ed. Belmont, CA: Wadsworth, 1979.
38. Montague, Peter. *Four Secure Landfills in New Jersey—A Study of the State of the Art in Shallow Burial Waste Disposal Technology* [PU/CEES 135].

Princeton, NJ: Princeton University, Center for Energy and Environmental Studies, 1982.
39. Montague, Peter. "Hazardous Waste Landfills: Some Lessons from New Jersey." *Civil Engineering/ASCE* (September 1982), pp. 53–56.
40. Moore, Charles A., and Elfatih M. Ali. "Permeability of Cracked Clay Liners." In *Land Disposal of Hazardous Waste; Proceedings of the Eighth Annual Research Symposium* [EPA–600/0–82–002]. Washington, DC: U.S. Government Printing Office, 1982, pp. 174–78.
41. Morrison, Allen. "Can Clay Liners Prevent Migration of Toxic Leachate?" *Civil Engineering-ASCE* (July 1981), pp. 60–63.
42. Mosher, Lawrence. "Who's Afraid of Hazardous Waste Dumps? Not Us, Says the Reagan Administration." *National Journal* (May 29, 1982), pp. 952–57.
43. Mossholder, Nelson V. "Statement of Nelson V. Mossholder, Technical Director of Stablex Corporation, before the Subcommittee on Natural Resources, Agriculture Research and Environment, [U.S.] House [of Representatives] Committee on Science and Technology, November 30, 1982." Unpublished. Available in typescript from the subcommittee.
44. Neely, N.; D. Gillespie; F. Schauf; and J. Walsh. *Remedial Actions at Hazardous Waste Sites; Survey and Case Studies* [EPA–430/9–81–05; SW–910]. Washington, DC: U.S. Environmental Protection Agency, 1981.
45. Nuclear Regulatory Commission. "Alternative Disposal Technologies." In *Draft Environmental Impact Statement on 10 CFR Part 61 "Licensing Requirements for Land Disposal of Radioactive Waste, Appendices A-F* [NUREG–0782, vol. 3]. Washington, DC: U.S. Government Printing Office, 1981, pp. F-1–F-97.
46. Palmer, Philip A. "Statement of Philip A. Palmer of the Du Pont Company on Behalf of the Chemical Manufacturers Association before the Subcommittee on Natural Resources, Agriculture Research and Environment, Committee on Science and Technology, United States House of Representatives, on Land Disposal of Hazardous Wastes. November 30, 1982." Unpublished. Available in typescript from the subcommittee.
47. "RCRA Regulations Rewrite Provokes Dispute on Efficacy and Raises Hackles on the Hill." *Chemical Marketing Reporter*, vol. 221 (January 11, 1982), p. 1.
48. Robertson, J. B. "Shallow Land Burial of Low-Level Radioactive Wastes in the United States—Geohydrologic and Nuclide Migration Studies." In International Atomic Energy Agency, *International Symposium on the Underground Disposal of Radioactive Wastes* [IAEA-SM-243/152] Washington, DC: U.S. Government Printing Office, 1981.
49. Robinson, H. Clay. "Testimony of H. Clay Robinson for the Hazardous Waste Treatment Council, to [U.S.] House [of Representatives] Subcommittee on Natural Resources, Agriculture, Research and Environment, [Committee on Science and Technology], November 30, 1982." Unpublished. Available in typescript from the subcommittee.
50. *Sanitary Landfill*. Manual no. 39. New York: American Society of Civil Engineers, 1976.
51. Sanjour, William. "Statement of William Sanjour, Chief, Hazardous Waste

Implementation Branch, U.S. Environmental Protection Agency, before the Subcommittee on Natural Resources, Agriculture Research and Environment, Committee on Science and Technology, [U.S.] House of Representatives, November 30, 1982." Unpublished. Available in typescript from the subcommittee.

52. Schultz, David W., and Michael P. Miklas, Jr. "Procedures for Installing Liner Systems." In *Land Disposal of Hazardous Waste; Proceedings of the Eighth Annual Research Symposium* [EPA–600/9–82–002]. Washington, DC: U.S. Government Printing Office, 1982, pp. 224–38.
53. Skinner, Peter N. "Facing the Chemical Waste Disposal Dilemma: The Newco Administrative Proceedings." *Capital University Law Review*, vol. 9 (Spring 1980), pp. 547–66.
54. Skinner, Peter N. "Performance Difficulties of 'Secure' Landfills for Chemical Waste and Available Mitigation Measures." In John P. Collins and Walter P. Saukin, eds., *The Hazardous Waste Dilemma: Issues and Solutions.* New York: American Society of Civil Engineers, 1981, pp. 32–56.
55. Stephens, Robert D., and Robert L. Judd. Correspondence dated July 15, 1982, to Lois Gibbs, Citizens Clearinghouse for Hazardous Wastes, Arlington, VA, re "Proposed Regulations to Prohibit Land Disposal of Toxic Wastes." This correspondence included a paper titled "Statement of Reasons for Proposed Regulations, 'Land Disposal Restrictions.' " Available from the authors at California State Department of Health Services, 714/744 P Street, Sacramento, CA 95814.
56. Stewart, Wilford S. *State-of-the-Art Study of Land Impoundment Techniques* [EPA–600/2–78–196]. Washington, DC: U.S. Government Printing Office, 1979.
57. Stoddard, S. Kent; Gary A. Davis; and Harry M. Freeman. *Alternatives to the Land Disposal of Hazardous Wastes; An Assessment for California.* Sacramento: [California] Office of Appropriate Technology, 1981.
58. Tolman, Andrew L.; Antonio P. Bellestero; and others. *Guidance Manual for Minimizing Pollution from Waste Disposal Sites* [EPA–600/2–78–142]. Washington, DC: U.S. Government Printing Office, 1980.
59. U.S. Environmental Protection Agency. "Draft RCRA Guidance Document, Landfill Design, Liner Systems and Final Cover [to Be Used with RCRA Regulations Sections 264.301(a) and 264.310(a)], Issued: 7/82." Available in typescript from U.S. Environmental Protection Agency, Land Disposal Branch, Office of Solid Waste, 401 M Street, SW, Washington, DC 10460.
60. U.S. Environmental Protection Agency. "40 CFR Part 264 [SW FRL 1626–4]. Standards Applicable to Owners and Operators of Hazardous Waste Treatment, Storage and Disposal Facilities." *Federal Register*, vol. 45 (October 8, 1980), pp. 66816–23.
61. U.S. Environmental Protection Agency. "Hazardous Waste and Consolidated Permit Regulations." *Federal Register*, vol. 45 (May 19, 1980), pp. 33066–588.
62. U.S. Environmental Protection Agency. "Hazardous Waste Management System. Addition of General Requirements for Treatment, Storage, and Disposal Facilities (40 CFR Part 164): Amendment of Interim Status Stan-

dards Respecting Closure and Post-closure Care and Financial Responsibility (40 CFR Part 165); and Conforming Amendments to the Permitting Requirements (40 CFR Part 122). *Federal Register*, vol. 46 (January 12, 1981), pp. 2802–97.

63. U.S. Environmental Protection Agency. "Hazardous Waste Management System; Permitting Requirements for Land Disposal Facilities." *Federal Register*, vol. 47 (July 26, 1982), pp. 32274–388.
64. U.S. Environmental Protection Agency. "Hazardous Waste Management System: Standards Applicable to Owners and Operators of Treatment, Storage and Disposal Facilities, and Permit Program." *Federal Register*, vol. 46 (February 5, 1981), pp. 11126–77.
65. U.S. Environmental Protection Agency. "Hazardous Wastes—Proposed Guidelines and Regulations and Proposal on Identification and Listing." *Federal Register*, vol. 54 (December 18, 1978), pp. 58946–9028.
66. U.S. Environmental Protection Agency. "Standards Applicable to Owners and Operators of Hazardous Waste Treatment, Storage and Disposal Facilities." *Federal Register*, vol. 46 (May 26, 1981), pp. 28314–28.
67. U.S. General Accounting Office. *Waste Disposal Practices—A Threat to Health and the Nation's Water Supply* [CED–78–120]. Washington, DC: U.S. Government Printing Office, 1978.
68. Ware, Sylvia A., and Gilbert S. Jackson. *Liners for Sanitary Landfills and Chemical and Hazardous Waste Disposal Sites* [EPA–600/9–79–005]. Washington, DC: U.S. Government Printing Office, 1978.
69. *Will County Disposal, Inc. Description of Design and Plan Operation for Secure Landfill "A" in Will County, Illinois*. Chicago: PRC Consoer Townsend, July 1982.
70. Worthy, Ward. "Hazardous Waste: Treatment Technology Grows." *Chemical & Engineering News* (March 8, 1982), pp. 10–16.

2

Information Burdens and Difficulties in Conceptualizing the Crisis: A Reappraisal

GARY N. DIETRICH

The history of the nation is filled with examples of American society discovering social problems, grasping for laws and institutional devices to address these problems, groping for knowledge to better understand the issues, making mid-course corrections as experience increases, and, finally, arriving at reasonable and workable solutions. The nation's discovery and response to the hazardous waste mismanagement of yesteryear is another chapter in this history, a chapter that is still being written.

In 1976, the Congress and the Environmental Protection Agency (EPA) had a vague idea that hazardous wastes were not being properly managed and could cause severe harm to the public health and the environment. With almost no public debate, Congress enacted the Resource Conservation and Recovery Act, known as RCRA. Subtitle *C* of RCRA mandated that the EPA develop a comprehensive regulatory program to control the management of the nation's hazardous waste. Thus, with an imperfect understanding of the dimensions of the problem and with little knowledge of how to actually manage hazardous wastes safely, the opening pages of a new chapter in history were written.

In early 1977, EPA began its task with the belief that the job was reasonably simple and could be accomplished in the 18-month time frame stipulated by the statute. A set of criteria would be developed to define hazardous wastes, compelling industry to identify their wastes. Also, a set of rules ensuring that hazardous wastes were sent to authorized hazardous waste management facilities would be developed. Finally, the experience that had been gained in the design and management of sanitary landfills and other refuse facilities would be extrapolated to develop rules for managing hazardous wastes. It was perceived to be a simple problem with a straightforward solution.

EPA began its task only to learn that the problem was far more complex than Congress first imagined. Moreover, it discovered that it lacked sufficient data and the scientific understanding necessary to construct reliable regulations and a workable program. Thus, what started out to be an 18-month task grew into a six-year odyssey during which the hazardous waste rules were promulgated in three major phases—in May 1980, January 1981, and July 1982. Meanwhile, the patience of the public and the Congress grew thin.

This six-year period was a learning period, during which EPA groped for the knowledge, data, and scientific principles necessary to understand the problem and frame a solution. Since monies and personnel for research and technical studies were distributed late and in small amounts, the Agency was often forced to conceptualize the real world of hazardous wastes. Thus, the groping-for-knowledge section of the chapter was written with great frustration and amid a shower of criticism.

Today, hazardous waste programs are only midway toward correcting the problems. A set of rules and a national program for hazardous waste management are in place. Despite frequent allegations to the contrary, the program is working, not perfectly, but a great deal better than some would have us believe. The nation is conscious of the ills of hazardous waste management, attitudes about hazardous waste management have vastly improved from the "bury it and forget it" mentality of yesteryear, and large quantities of hazardous wastes are, today, being responsibly managed.

Yet, there is more to be done. The question of whether land disposal of hazardous waste should be severely restricted is before us. The issue of whether the hazardous wastes produced by small-quantity generators should be more extensively regulated remains. The matter of regulating the burning of hazardous wastes as fuel is up for consideration. These matters are currently being debated by the Congress and being examined by EPA, and properly so, because they need to be resolved. Thus, the nation is presently engaged in writing the third and probably last segment of the chapter on hazardous wastes.

As we move into the midpoint of the learning campaign, it is useful to step back and examine where we have been, where we are, and where we are going. The following text attempts to do that with respect to three issues that have been and continue to be central to the nation's hazardous waste program.

WHAT IS A WASTE?

In the early days of hazardous waste rule making, the issue of what is a waste did not seem difficult. After all, a waste is a material that is discarded; so what is the definitional problem? Then, it was recognized that the spreading of dioxin-contaminated waste oil for dust control at a horse arena in Verona, Missouri, had caused some disastrous effects: the loss of many animals and the severe illness and near loss of a child. (This situation ultimately led to the dioxin contamination at Times Beach, Missouri, which has more recently been the focus of a controversial Superfund cleanup activity.) This was a situation where a material—waste oil—was not being discarded, but instead, was being reused. Then the Silresum site near Lowell, Massachusetts, was discovered. A large quantity of various toxic chemicals were being held in inventory for subsequent resource recovery. Unfortunately, because of technical and financial difficulties, the processing of these chemicals was not keeping pace with incoming receipts. As a result, the inventory in storage grew and grew, and, because of poor housekeeping procedures, the chemicals began to leak from their containers causing environmental damage. In addition, this collection of wastes threatened the air and constituted a fire hazard, since a large portion of the chemicals were highly volatile, flammable, and toxic. Here was a situation where materials were not being discarded; instead, they were stored for recycling and reclamation.

Many similar situations were discovered by EPA inventories. Spent materials—such as used oils and solvents, or manufacturing by-products, were being used, reused, recycled, or reclaimed in a manner that posed a hazard to public health or the environment. On the other hand, there were countless situations where such practices were being performed with little or no adverse environmental effects. Thus, the dilemma: Should certain materials that are waste-like but that are not discarded be defined and regulated as wastes? Should this be done to control the unsafe practices exemplified above, even though it placed regulatory requirements on other common practices that were safely performed and did not justify control?

EPA struggled with this issue. Ultimately, it concluded that RCRA gave the EPA regulatory jurisdiction over waste-like materials that are not, in fact, discarded. It further concluded that the use, reuse, recycling, and reclamation of waste-like materials, in certain cases, needed to be and should be regulated. But the drafting of such regulatory coverage was quickly mired in the complexity of the subject matter. As in so many other areas of the hazardous waste program, EPA had virtually no data on the thousands of waste-like materials (that is, spent materials and manufacturing by-products) being produced; the degree or potential degree to which these materials were used, reused, recycled, or reclaimed; the adverse environmental consequences, if any, of such practices; or the results that might follow their regulation. There was insufficient time, due to court-ordered schedules, and virtually no funds or personnel to build a data base. After extensive meetings and discussions, the EPA did the best that could be done: apply professional judgment to conceptualize the actual situation and develop a rule that would fit.

Thus, the EPA experienced firsthand all the debilitating effects that insufficient data inflicts on rule making. With a large base of factual data, it is easy to develop a reasonable, workable, and effective rule; to adjust that rule, if appropriate, based on new data and public comments; and to defend that rule, if necessary, in the courts. In the absence of sufficient data, however, the regulatory agency is left to conceptualize where the facts might lie and to build a rule based on such conceptualization. When objections come forth in public comments or in litigation, the agency is naked and often unable to defend its conclusions convincingly. It is easily found to be arbitrary, and maybe capricious, in its rule making and, if so found by a court of law, the rule will be overturned, stayed, or remanded to the agency for reconsideration.

In the proposed RCRA hazardous waste regulations published in December 1978, EPA proposed a definition of wastes based on its best judgment. This drew severe criticism, but these comments yielded virtually no data to help the agency in its deliberations. EPA, again under court orders, developed a revised definition of wastes, which it published in May of 1980. This revised rule also drew criticism and was one of the issues for which petitioners sought judicial review.

Both EPA and the petitioners sought to negotiate this issue before bringing it into the courtroom. As a result, over the last year and a half, some 30 negotiation sessions have been held, and endless hours have been spent by EPA to revise the definition of waste. As of this writing, EPA is near proposal of a very complicated but much improved revision of the definition, which again must go through public comments before it can take effect.

Much of the current debate over the efficacy of the existing hazardous waste programs hinges on an effective definition of toxic waste. Foremost among these is the issue of regulating the burning of hazardous wastes in boilers. The burning of wastes or waste-like materials in a boiler or other burning devices for the capture and use of their fuel value is actually a use or reuse of such materials, currently exempt under RCRA. If RCRA jurisdiction is to reach these burning practices, then the definition of wastes must include waste-like materials. Similarly, the issue of regulating the use of used oils for road oiling and other purposes depends on an apt definition of waste. Even the regulation of the reclamation of spent solvents, designed to avoid Silresum-type situations, depends on appropriate definition of wastes.

By late in 1983, a general consensus had developed that these and certain other practices are justifiably regulated under RCRA. If EPA does not issue its own re-definition of waste, it is likely that Congress will demand the refinement soon. Even so, to the student of environmental regulation, the next policy steps on the definitions of wastes have profound implications. Once the barrier is broken of restricting wastes to those materials that, in fact, are discarded, then RCRA jurisdiction can creep into the manufacturing process and into other areas of regulation. Some believe this is the province of the Toxic Substances Control Act. For example, if off-site reclamation of spent solvents is to be regulated by RCRA, then why not also regulate on-site reclamation of these materials? In recent years, it has become increasingly common for solvent recovery to become an integral part of the manufacturing process both because of the increasing cost of virgin solvents and the regulation of volatile organic compound emissions by air regulations. In the textile and dry-cleaning industries, for instance, it is not uncommon to collect spent solvents and solvent vapors that once were discarded and then to clean and recycle them in a closed-loop system built right into the industrial process. Many would say that these and other recycling processes, if they need to be regulated, are not best handled by RCRA. But, if caution is not exercised, RCRA jurisdiction will find its way into these processes and deep into the manufacturing process.

As noted above, the paucity of a comprehensive data base severely impeded the early resolution of the definition of wastes. In addition, the inertia of strongly held opinions also interfered with its resolution. Initially, the regulated community stood fast in their contention that a material had to be discarded, or at least unwanted, to be a waste. Gradually, as the shortcomings of such a restricted definition came to be known and appreciated, the literal interpreters began to moderate their views and accept a broader definition of the term.

This gradual change of thinking, and its related overcoming of inertia, represents the recurrent patterns of reform experienced by the EPA. Toxic waste rule making is not a simple, straightforward process. Indeed, since 1976 rule-making in this field has been accompanied by an education process: one that has gradually changed the old ideas, concepts, and attitudes of the public, the regulated community, the EPA, and the Congress. Change doesn't occur overnight or by virtue of a new statute. Change only occurs by changing attitudes, a painfully slow and tedious process.

WHAT IS A HAZARDOUS WASTE?

Concurrent with wrestling with the question of what is a waste, EPA began, in 1976, to face the more difficult question: What is a hazardous waste? The statute clearly mandated the EPA to identify a subset of wastes that presented "substantial present or potential hazard to human health or the environment." Since all wastes, even household garbage, are hazardous to some degree, this was a tall order—one that might better have been delegated to King Solomon.

EPA next engaged in the fine art of demarcating hazardous from non-hazardous wastes. A set of physical, chemical, and biological characteristics were used. EPA requested that waste generators assess their wastes according to these guidelines. This was a fine idea but one that ran into real-life problems as the details of such a system began to be worked out.

The first and most significant problem involved test methods to implement this scheme. What tests were widely available so that the regulated community, represented by an astonishingly diverse array of experiences, could use the same tests competitively? With respect to ignitability, reactivity, corrosivity, and toxicity, the EPA felt that it could require workable uniform tests. Relatively simple, inexpensive, widely used equipment for standardized testing was generally available for these characteristics. In addition, it was concluded that most generators of waste (with the exception of small generators of wastes) could reasonably be expected to implement these test methods either themselves or through commercial testing laboratories.

Nevertheless, there remained, and to some degree still remains, several problems: How do you ensure that a representative sample of the waste is being tested? How frequently do you need to test the waste? Are the results of the toxicity test reproducible? Is the toxicity test representative of the waste properties it is designed to measure? And what do you do when you run into a waste that interferes with the testing apparatus or test method? These subsidiary problems were resolved by 1983, but it has taken time and a great deal of research.

Yet once again we are only at the midpoint. EPA has not yet performed a comprehensive investigation of how carefully and thoroughly the tests are performed for a wide spectrum of non-listed wastes. It would not be surprising to learn that, for these wastes, the tests are working poorly. When one goes into the field and observes the great variety of wastes generated by thousands of manufacturing processes, sees how wastes are often mixed and intermingled together, and, finally, achieves a sense of the billions of pounds of the wastes generated yearly in America, one comes away with the impression that it would be extremely unlikely that the characteristic approach to hazardous waste identification is catching all that it is meant to catch. Even with a well-meaning and environmentally alert regulated community, it is doubtful, in the absence of enormous expense, that all wastes can be sampled with sufficient frequency and thoroughness. This is an ugly fact that must be factored into all rule making. What is probably helping to make the current approach work is that many wastes are managed as hazardous by generators to avoid liability, as soon as there is any evidence that they sometimes exhibit any of the legislated characteristics.

Another problem with the characteristic approach was non-representativeness. During rule making, it was alleged that some common commercial products, and even foodstuffs, would exhibit the characteristics of hazardous wastes. The leading allegation was that although tomato skins having pesticide residues are well within FDA tolerances, they still exhibit the proposed characteristics of hazardous wastes. Another was that newspapers would also qualify as hazardous wastes. Although EPA had answers to these allegations, regulators feared that an example demonstrating a fault in the characteristic approach could crop up to undermine the entire effort. After all, EPA had seen the power of several interest groups that, after seeing the proposed regulations of December 1978, were successful in securing temporary exemptions from RCRA jurisdiction for mining wastes, oil and gas production wastes, and utility wastes. While the public and the Congress might like to think that the scientific and technical conclusions would prevail in situations like those described above, the truth is that EPA had to keep its eye on the political perception of the hazardous wastes rules it was writing. Any flaws in a rule could be quickly taken out of context and magnified by special interest groups to overturn the rule.

As problems with the initial characteristic approach began to surface, the EPA considered a listing approach, which would itemize specific wastes as hazardous wastes. This approach was deemed to have the advantages of providing certainty to the regulated community, states, and the EPA for the enforcement of regulations. It would not be necessary to test wastes—to apply the characteristics on a case-by-case basis—but only to ascertain whether the waste met the listing description. But this approach also carried a significant disadvantage. It placed the burden of proof on EPA to discover and identify wastes, a burden that depended on extensive resources. Industry generally liked this approach. Environmental interest groups disliked this approach because they did not believe that EPA would have the resources to adequately carry the burden and, even if it did, the process would be time consuming. In the end, EPA was faced with the difficult choice of whether to employ the characteristic approach for toxic organic wastes or the listing approach. In the May 1980 regulations, it chose the second approach and listed several hundred hazardous wastes.

The task of discovering, characterizing, and listing hazardous wastes is an enormous one given the high variety of hazardous wastes generated by our chemical society. EPA has a program committed to this effort, but funding has been modest. Moreover, this task is exhausting if done in the traditional way of listing each individually different waste stream. The waste produced by a particular process can be varied significantly by slight changes in operating temperature, reaction time, pressure, catalyst, or even small variations in raw materials. These operating conditions may change frequently as technology changes, markets change, or product improvements are made. To bring the scope of listing these wastes within reasonable hope of accomplishment in a timely manner and with the resources available, the EPA has been developing generic listings in which hundreds of slightly different wastes within "a product/process category" are included in one regulatory listing. As of this writing, the first listing under this approach has yet to be proposed, but EPA is near such a proposal. If it works, that is, if this approach can withstand the criticism that industry will undoubtably raise, it will provide a more orderly and efficient way of listing the hazardous organic wastes that are notably absent from the current scheme.

Without a doubt, the RCRA hazardous waste program is probably a long way from covering all of the hazardous wastes that should be regulated. The choice of the appropriate approach to correct this deficiency is not an easy one. Probably the listing approach with the liberal use of generic listing will prove the most certain, comprehensive and efficient. Yet this requires not only that EPA be given the resources it needs, but also a clear mandate to proceed down this pathway openly and promptly.

3

Keeping the EPA Vigilant: The Role of Private Watchdog Agencies

KHRISTINE L. HALL

Citizen groups, both national and local, have played a significant role in ensuring proper enforcement of the nation's hazardous waste laws. In passing the environmental statutes of the 1970s, Congress realized that state and federal agencies alone could not perform the monumental task of cleaning up the environment and preventing further degradation. As a result, Congress subjected state and federal efforts to public scrutiny and created numerous avenues for public input. This is especially important for agencies governing hazardous waste. In no other arena have citizen groups shown themselves to be so necessary to ensuring protection of public health and the efficacy of government programs than in this most controversial part of toxic waste controls.

The Environmental Defense Fund (EDF), a national nonprofit environmental organization, combines the skills of lawyers and scientists to oversee the efforts of state and federal agencies. A key concern is the implementation of statutes to control hazardous waste. Citizen groups like EDF have a number of important roles in the regulation of hazardous waste. Of primary importance is the function of informing the general public of regulatory action, inaction, or malfeasance. Polls consistently show very strong support for environmental regulations; hazardous waste ranks at the top of these concerns expressed by the public. Yet because of the arcane and detailed nature of many agency decisions, interested members of the public are often unaware of the day-to-day decisions that can affect their lives. Once alerted, many people have been willing to translate their concern into action, but it is the role of these private watchdog agencies like EDF to serve as the translator. Methods of alerting range from newsletters, articles, and publications to direct work with the press.

Citizen groups also provide critical information directly to agencies during their decision-making processes. Creative approaches and technical information are often part of presentations by citizen groups. This alternative point of view on technical and legal matters often acts as a necessary counterbalance to the input supplied to the EPA by industry and the regulated community. As acknowledged by the courts, governmental rule making works best when it is weighted by influences from diverse directions.

Another role of citizen groups is to force a reluctant agency to act. When Congress mandates the EPA to publish standards, that is usually only the first step of an elaborate process. Getting EPA to carry out Congress' directives is often quite another story. For a variety of reasons—budget restrictions, bureaucratic inertia, a tendency to put off tough decisions, basic disagreement with the congressional mandate—EPA has often stalled or inadequately carried out its statutory requirements. Although Congress' oversight authority is a powerful tool, as witnessed by the recent rounds of resignations at EPA, it is often fairly limited in application. Congress is usually too busy to scrutinize the details of agency actions in fulfilling the statutes it writes. Congress has therefore given citizen groups the authority to take legal action to force EPA compliance with statutory requirements. Often this authority is used in "deadlines litigation"—lawsuits to force EPA (under court order) to promulgate regulations by a certain date. In other instances, citizen groups use this authority to challenge the substance of an EPA rule.

Lastly, citizen groups have the authority to act as private attorneys general, serving as a surrogate government enforcement official, by suing those who violate EPA regulations. Congress intended this to supplement enforcement efforts of state and federal governments. Unfortunately, these private enforcement actions are infrequent. Information identifying violations is not easy to acquire. Monitoring reports documenting violations are usually located in regional offices many miles from an offending facility, where citizens do not even know of their existence. In addition, litigation is cumbersome, expensive, and complicated. Many local attorneys are not yet familiar with the technical or legal details of hazardous waste law.

Two recent examples of action by citizen groups prove their indispensable role in establishing reliable hazardous waste controls: (1) EPA's delayed publication of regulations controlling land disposal facilities and (2) EPA's suspension of its ban against dumping drums containing liquid hazardous waste.

LAND DISPOSAL REGULATIONS

The most pressing hazardous waste problem facing the nation is the degree to which we dispose of waste on land. Approximately 80 percent of hazardous waste is dumped on land—the least desirable means of disposal. Yet, it was not until July 1982—almost six years after Congress called for efforts to upgrade dumps—that EPA got around to writing the basic regulations to control landfilling of toxic wastes. What is more alarming is that even this delayed schedule would not have been met had it not been for the efforts of environmentalists to force EPA to carry out its statutory duty.

In October of 1976, Congress enacted the Resource Conservation and Recovery Act (RCRA). Subtitle *C* of RCRA created a regulatory program to improve the nation's hazardous waste management practices. To accomplish that purpose, Congress required EPA to develop a set of regulations that designated which industrial wastes are hazardous; specified procedures and standards for generators and transporters of hazardous waste; set permitting procedures and standards for treatment, storage, and disposal facilities; and set requirements that the states must meet to run the RCRA program in lieu of EPA. In response to the urgency of this country's hazardous waste problem, Congress mandated the publication of these regulations within 18 months, or by April 21, 1978.

The statutory deadline came and went and EPA had not even proposed the required regulations. Since the Agency was not devoting sufficient resources toward implementing the program, further prolonged delays seemed inevitable.

In response to this standstill, the EDF and several other organizations filed suit in September of 1978 to force EPA to do its job, using the citizen suit provision in RCRA. On November 7, 1978, in response to an EDF request, the court found EPA had violated RCRA by not adhering to its statutory mandate.

EPA then proposed and the court adopted a schedule that required the agency to issue all the regulations by the end of 1979. Unfortunately, however, EPA failed to meet its own schedule. EPA then proposed a second schedule, which the court again adopted. This schedule required that regulations governing generators and transporters of hazardous waste as well as state requirements would be published in February and April of 1980. It also required that final standards covering treatment, storage, and disposal facilities—the landfills, impoundments, and incinerators where waste ends up—would be promulgated by the fall of 1980.

The efforts of the Environmental Defense Fund and the court began to bring results in 1980 and early 1981. EPA met most of its second schedule, except for final standards governing land disposal facilities and a few other minor facilities.

The Environmental Protection Agency had, however, essentially left the toughest problem—and the most dangerous one—to be considered last. Land disposal facilities include such things as landfills, surface impoundments (pits, ponds, and lagoons full of hazardous waste), waste piles, and land treatment of hazardous waste. These facilities have a long history of being built with little or no consideration for the leakage of poisons into groundwater or the air pollution problems caused by the chemicals. Evidence showed that surface impoundments had been deliberately built over sandy soils so that the materials could infiltrate the soils and be washed away. A report to Congress by the Library of Congress described 50 pollution incidents arising from inadequate land disposal practices, resulting in hundreds of well closings. Another report described the sobering results of an EPA assessment of pits, ponds, and lagoons: approximately 30% of the industrial facilities are unlined and sit directly over a groundwater source. Worst yet, about 10 percent are unlined, sit directly over an aquifer, and are within one mile of a potential water source. Delays in publishing regulations controlling these dangerous practices clearly represented a serious and continuing threat to public health and the environment.

Despite the urgency of the problem, EPA finally published landfill regulations in draft form, in October of 1980. On December 8, 1980, EPA Administrator Douglas Costle informed the court that the land disposal standards would be published in January 1981, or at some unspecified future date depending upon the need to redraft the regulations for public comment. Mr. Costle pledged to inform the court as to the Agency's intentions in January 1981.

On February 5, 1981, after Costle had left office with the rest of the Carter administration, EPA redrafted the land disposal regulations and asked for public comment, providing a 180-day comment period.

On May 26, 1981, EPA published a supplemental rule-making notice, requesting comments on a variety of regulatory approaches and further technical information the Agency felt were necessary to formulate regulations. It was clear from this notice that the new Reagan EPA was thoroughly reviewing past approaches and decisions and would not have a definite approach to landfill regulations for quite some time.

In the midst of this apparent confusion on how to approach land disposal regulations, EPA had not informed the court of their new schedule. EPA was, therefore, operating without the threat of a court-ordered deadline to hasten their actions.

In light of this, on August 13, 1981, EDF requested the court to hold a hearing to review the progress made to date. EDF had learned that the newly appointed EPA administrator, Anne M. Gorsuch, had approved an internal working schedule that pushed the date for the land disposal standards back to December 1982 at the earliest. In October 1981, the Environmental Protection Agency responded to EDF's request for an announcement of intent. In this affidavit, Administrator Gorsuch stated that the reasonable target date for the publication was two years away, or the fall of 1983. This target was five and a half years after the statutory deadline, four years after EPA's first proposed target date, three years after the Agency's second proposed target date, and two years after the close of the comment period on the most recent draft regulations.

EPA sought to justify this long period of delay by stating that the issues were complex, as evidenced by the responses received from the May request for information. Also, EPA stated that additional time was needed in order to prepare the regulatory impact assessment recently required by an executive order from President Reagan. This regulatory impact assessment essentially required EPA to conduct a cost-benefit analysis of the regulations, a requirement that is neither provided nor allowed by the governing statute. (See chapter 5 for more detailed evidence on the role of the Office of Management and Budget [OMB] in toxic waste rule making.)

On November 13, 1981, the District Court ordered EPA to promulgate the land disposal standards by February 1, 1982. The court was clearly fed up with EPA's delay tactics:

> A further extension is not warranted. In her report, the Administrator states that regulations governing land disposal of hazardous wastes raised technical and policy issues of great complexity and the Agency's prior research and analysis, conducted for the most part under the supervision of the prior administrator, do not provide an adequate basis for resolving these issues. The fact that RCRA directs the Administrator to promulgate regulations within 18 months of the Act indicates that Congress did not direct the Agency to resolve every conceivable problem before issuing regulations. In the more than 5 years since RCRA was enacted the Agency has had the opportunity to consider various regulatory approaches and numerous comments and suggestions by the affected interests.

On December 3, 1981, EPA requested the court to reconsider its order, restating its desire for a two-year delay and offering no alternative dates to February 1. The court denied the motion on December 7. In the meantime, the EPA's Office of Solid Waste, through its Director Gary Dietrich, initiated efforts to comply with the November 13 order, calling a public hearing on December 21, 1981, to solicit comments on its regulatory approach. Mr. Dietrich stated: "We are going to make a valid attempt to try to produce, by February 1, a responsible standard. I think we can do this."

Nevertheless, on January 7, 1982, EPA filed a notice of appeal in the Court of Appeals challenging the District Court's order to publish the regulations by February 1, 1982. On January 25, EPA again requested the District Court to reconsider its November 13 order, or at least to postpone its effect. Attached to the motion was an affidavit from Mr. Dietrich stating that the February 1 deadline was impossible to meet, a direct contradiction of his statement a month earlier. EPA did not see fit, however, to propose a new date by which they could complete the regulations.

EDF responded by taking Mr. Dietrich's deposition, in an effort to determine exactly how much time EPA thought necessary to issue the regulations. From this deposition, EDF learned that Mr. Dietrich felt the regulations could be finished by mid-March 1982.

Yet by exercising its legal right of appeal, EPA evaded another court-ordered deadline and bought another six months of delay. Briefs were filed by both parties during the spring of 1982 and the Court of Appeals heard oral argument on the case on June 1, 1982. At no time during this appeal process did EPA volunteer a date by which it could complete the regulations, even though they were prodded to do so by the Court of Appeals.

In mid-June 1982, the Court of Appeals decided to uphold the District Court's order. Because the February 1 deadline had long since passed, however, the Court of Appeals set a new deadline of July 16, 1982, for promulgation of the regulations. The regulations were finally published on July 26, 1982.

This four-year battle to compel EPA to do its statutory duty was, however, only the end of the first chapter. In the fall of 1982, the Environmental Defense Fund filed another lawsuit challenging the legality of the regulations published in July, on the grounds that they were impermissibly weak and did not protect public health.

The effort to force EPA to publish regulations governing the most dangerous method of waste disposal took over four years of time, a considerable legal expenditure. These critical regulations would have been delayed time and again, perhaps retarded forever in select cases, except for counter legal action. It is too often the case that an Agency such as the Environmental Protection Agency simply ignores a statutory deadline set forth by Congress; sometimes out of lack of appreciation of the urgency of the problem, sometimes out of lack of resources, and sometimes because the Agency just does not believe in what Congress has directed them to do. Since the sanction of contempt of Congress is exercised only in the most egregious situations and congressional oversight does not compel mandatory action, the role of citizen lawsuits in requiring agencies to take action is indispensable. These "deadline" suits, represented by the above case, have become an important part of hazardous waste regulation. Without citizen groups to fight for these deadlines, numerous regulations would be dangerously delayed.

SUSPENSION OF THE BAN ON DRUMMED LIQUID HAZARDOUS WASTES IN LANDFILLS

The Reagan administration took office in January 1981 with the perceived mandate to relieve the regulatory burden on industry. One of the main areas the new administration chose to concentrate on was environmental regulations. In carrying out its perceived mandate, the new Reagan EPA acted with a special vengeance on hazardous waste matters. At the time, rising hazardous waste regulations were either still being formulated or were in their infancy. In most cases, these regulations would add substantial new compliance costs to industry. Because of this, the hazardous waste regulations became a prime target for review and weakening.

One of the first requirements targeted by the chemical and waste disposal industries for review was a ban against the dumping of drums containing toxic liquids. This requirement had been published in May of 1980. But it was not to take effect for 18 months, to give industry time to find alternative methods of disposal. November 19, 1981, then, was to be the start of the ban.

The reasons for the ban were simple. Liquids rust drums faster than solids. Metal drums do not have a very long lifetime in landfills and soon start leaking their contents into the landfill. In addition, once a drum full of hazardous waste starts to leak, it becomes crushed down by the weight of other layers of waste on top of it, thus causing subsidence of the landfill. This subsidence causes cracks in the cap put on top of landfills to prevent infiltration of precipitation. When cracks are opened, precipitation enters the landfill, mixes with the hazardous chemicals, and causes the generation and eventual escape of even more hazardous chemicals into groundwater. The ban required the chemical industry to solidify its toxic waste or dispose of it through some other method besides landfilling, such as incineration.

After the Reagan EPA came in, industry started a court case attacking the ban against dumping drums of toxic liquid waste. Both the National Solid Waste Management Association (NSWMA), the trade group representing owners and operators of hazardous waste disposal facilities, and the Chemical Manufacturers Association, the trade group representing generators of chemical waste, asserted that the ban was impractical and unenforceable. After sustained negotiations, industry persuaded the EPA to modify its ban.

The compromise approach advocated by industry was much different from the one advocated by the Environmental Protection Agency. In October 1981, the National Solid Waste Management Association stated that the ban should be rescinded and that landfills should accept liquid hazardous waste for up to 25 percent of the entire volume of the landfill. In response, EPA proposed that the ban be modified to allow each individual barrel to contain up to 10 percent of liquid hazardous waste. The difference between the two approaches consists not only of the percentage of allowable liquid hazardous waste, but also the form that percentage is to take. EPA's 10-percent barrel formulation would preserve, to some extent, the integrity of the crucial cap over the landfill. By contrast, NSWMA's proposal would allow barrels to contain 100 percent liquid waste and thus would not ensure the integrity of the cap.

In the fall of 1982, industry presented its proposal to the EPA in a meeting where all litigants were present, including EDF. EPA regulators responded by outlining their proposal and then asked all parties to leave the room. After 20 minutes of deliberation, EPA called the parties back into the meeting room and announced the industry position had been accepted: 25 percent of a landfill could contain toxic liquids.

Since the ban had been formally promulgated as a regulation in May 1980, it was illegal for EPA to modify the ban according to the whim of industry. Administrative law requires that the Environmental Protection Agency go through rule-making proceedings (that is, publication of a proposed rule or revision thereof in the *Federal Register*, with the opportunity for the public to comment on it). Since the effective date of the ban, November 19, was close at hand, EPA promised industry litigants that it would temporarily suspend the ban and then go through the rule-making procedure by proposing industry's 25-percent approach for public comment.

The date to implement the ban came and went, yet there was no announcement by the EPA on the issue. The silence was puzzling. Some said that the necessary papers to take the action promised to the industry litigants had been lost in the office of Anne M. Gorsuch. Finally, in mid-February 1982, almost three months after the ban had already been in effect, environmentalists received word that EPA would soon be publishing its long-awaited action in the *Federal Register*.

Alerted to the impending EPA action, a coalition formed between environmentalists and representatives from the hazardous waste treatment industries that employed higher technologies. This may be the first time industry and environmentalists were so completely like-minded both in their timing and common approach for stronger regulations. The environmentalists, led by the Environmental Defense Fund, were concerned about the adverse public health and environmental consequences of EPA's actions. The hazardous waste managers, represented by the newly formed Hazardous Waste Treatment Council (see chapter 11 on HWTC), were concerned primarily about the economic effect of EPA's decision on their businesses. New markets had been formed by EPA's original rule, since a ban on drummed liquid in landfills would force liquid waste out of landfills resulting in more incineration of hazardous waste. The lifting of the ban would undercut this market and discourage capital investment in alternatives to landfills over the long run. The combination of the health and environmental concerns with economic concerns proved to be a potent force.

This coalition began to alert the public about the consequences of EPA's decision. EDF and the Hazardous Waste Treatment Council approached key media contacts. This unusual union between environmental and business interests, compounded by the outrageous actions of the EPA, provided the necessary "news hook" for widespread press coverage of the issue. Indeed, newspapers across the country featured front page coverage and frequent editorials on the subject. Radio and television coverage were also widespread. The result was an upswelling of public outcry and opposition, expressed directly in letters to EPA or to congressional representatives. Adverse congressional reaction was also swiftly forthcoming, as Senators and Representatives from both sides of the aisle publicly castigated or privately put pressure on EPA.

As a second line of attack, environmentalists and HWTC also filed lawsuits, calling for the courts to review EPA's decision. The parties alleged that the Environmental Protection Agency had acted unlawfully in violating both the Resource Conservation and Recovery Act and the Administrative Procedures Act.

As a result of ths double punch of public outcry and legal action, the Environmental Protection Agency reversed its position within a month, admitted its mistake and reinstated the ban on the disposal of drummed liquid waste in landfills. Although this episode ultimately had a happy ending, the fact remains that in the month during which the ban had been unlawfully lifted, tens of thousands of drums of liquid hazardous waste were disposed of in landfills with impunity. The ban would not have been reimposed so quickly had it not been for the efforts of citizen groups and the extraordinary public response.

It is always dismaying when a government agency is forced by the public to obey the law. The two episodes illustrate this now common occurrence with the EPA. But in the case of toxic wastes, there are additional reasons for dismay. Federal delays in passing toxic waste controls represent the most insidious form of deregulation. Without the tenacity of private watchdogs to force the EPA to strengthen its bite, unsafe dumping will continue and safer treatment alternatives will remain stifled.

4

Dumping into Surface Waters: The Making and Demise of Toxic Discharge Regulations

JAMES BANKS

"It is the national policy that the discharge of toxic pollutants in toxic amounts be prohibited."[1] With that pronouncement in the opening paragraph of the 1972 Clean Water Act, Congress initiated one of the most intricate policy debates in modern environmental law. Attorneys, scientists, and lawmakers have struggled for over a decade to create a rational regulatory program for implementing this national policy.

The stakes are well worth the struggle. Every day, American industry discharges over a million pounds of toxic chemicals into the sewers and surface waters of this country.[2] No one really knows how much of this chemical soup remains in the water, the sediments, or the tissues of fish and shellfish consumed by the public. Scientists are unable to predict the health effects or economic consequences of using America's waterways as industrial sewers. But few would disagree with Congress' conclusion that toxic pollutants must be curtailed. Consider only the examples that have made headlines:

- It was discovered a few years ago that kepone had produced severe neurological disorders in workers at the Life Science Products facility in Hopewell, Virginia, where the pesticide was manufactured. From that facility, many tons of kepone were dumped into the James River and Chesapeake Bay. Experiments on sheepshead minnows, a common species in the Chesapeake, showed that when exposed to nearly the lowest detectable level of kepone, the fish developed convulsions, and 100 percent of the fish developed neurological damage and the "broken back syndrome."[3] The minnows actually splintered their own backbones because of the convulsions.
- PCBs, a material used in transformers, plastics, paints, and lubricants, have found their way into the aquatic environment via industrial spills, discharges,

and leakage from discarded products. General Electric plants along the Hudson River have discharged as much as 30 pounds of PCBs per day into the river, used by over 100,000 people for drinking water. Fish downstream from the GE plants have concentrated PCBs in their tissues to an average level of 100 parts per million (ppm)—many times the FDA limit for edible fish.[4] In Lake Michigan, trout have been found with concentrations of 145 ppm.[5] Moreover, since the *average* concentration of PCBs in human breast milk is 1.8 ppm, the EPA reports that nursing infants are getting 10 times the safe level.[6]

- Thus far, investigators have found 112 organic chemicals in New Orleans' drinking water, including several known and suspected carcinogens. They expect the total may reach 900 chemicals after further study. EPA has concluded that Mississippi River drinking water is dangerous, especially for sick and elderly persons and for children. The rate of cancer mortality in New Orleans is 32 percent higher than the national average, and it has been estimated that polluted drinking water may account for 10 to 20 percent of the city's cancer death rate.[7]
- Chloroform, a carcinogen, is present in Miami's water at 311 parts per billion (ppb), the highest level in the nation.[8] Because this chemical concentrates in the human body, fetuses have been found to have higher concentrations than their mothers. And scientists have demonstrated positive correlations between chloroform levels in water and bladder and rectal cancers.
- In February 1977, a 70-ton slick of carbon tetrachloride, a carcinogen, stretched 100 miles along the Ohio River downstream of the FMC Corporation plant near Charleston. The slick was discovered after contamination levels had reached 130 ppb in *treated* drinking water in Huntington, West Virginia.
- Later that year, the large sewage treatment plant in Louisville, Kentucky, was knocked out of operation by two chemicals (hexachlorocyclopentadiene and octachlorocyclopentine) which had disrupted the normal degradation processes of the plant. Consequently, about 100 million gallons of raw sewage were discharged by the city into the Ohio River each day, and the dissolved oxygen levels in the river dropped substantially, producing the threat of a major fish kill. The decontamination effort, requiring special equipment to protect workers from dangerous fumes, cost over $4 million.

Similar accounts could be listed for each state. Rivers, lakes, and coastal areas are being closed to fishing and swimming. Commercial fisheries are frequently closed because of contaminated or malformed fish.

This kind of "evidence" has convinced Congress and the public that we have a toxic waste problem in America's waterways. And while examples might not establish the extent or severity of toxic contamination, such incidents have moved the nation toward an ambitious program to deal with chemical discharges.

Today's toxic waste control effort is a mixture of policies. The first line of defense consists of uniform federal regulations that require the best available technologies for control of discharges. These controls are backed up by ambient water quality standards as well as health-based standards, derived in the laboratory, for a few specific toxic chemicals.

These policies have evolved through 12 sessions of Congress, four administrations, and exhaustive litigation in the federal courts. Indeed, the overall control strategy may be unique among environmental programs in the degree to which environmental lawyers and the courts have directed the details of its evolution and implementation. For seven years, the Natural Resources Defense Council (NRDC) has had EPA under a federal court order that prescribes all of the Agency's responsibilities and deadlines for developing toxic discharge controls.[9] Known as the "Flannery Decree" (after U.S. District Judge Thomas Flannery who enforced it), the order not only changed federal policy dramatically in 1976, but heavily influenced amendments to the Clean Water Act that have guided EPA's efforts since 1977.

With the advent of the Reagan administration in 1981, EPA's comprehensive, integrated program has been under attack. The courts have repelled EPA's and industry's efforts to undermine the Flannery Decree, but the battle ultimately will be fought on Capitol Hill, where lawmakers are under pressure to extend deadlines, create variances, and dilute EPA's authority. With a decade of experience and considerable hindsight, Congress has to decide whether this program should proceed as planned. On that decision rides the next decade's aquatic environment.

THE GENESIS OF TOXIC WATER CONTROL

How much knowledge is required before a regulator should impose controls on the use and disposal of chemicals? Like the FDA and OSHA, EPA grapples with this question under all of its environmental statutes, and the Clean Water Act is no exception. When Congress established the Act's national toxics policy, several critical questions were left for EPA to work out: What *is* a toxic pollutant? *How much* of each chemical constitutes a "toxic amount" in the aquatic environment? And how should EPA limit discharges without going too far or stopping short of adequate health protection?

In 1972, Congress chose an ambitious approach. Under a special section of the Act, EPA would derive a list of chemicals it considered "toxic" and then within six months develop effluent standards that established allowable discharge levels for each substance.[10] The standards would incorporate an "ample margin of safety" to account for scientific uncertainty. Every industrial source of those pollutants would have to meet EPA's standards within one year, regardless of the economic hardship or the existing quality of water.

This listing approach differed dramatically from the scheme Congress chose for more conventional pollutants, such as degradable materials that rob the water of dissolved oxygen needed to sustain aquatic life. For these more common and scientifically easier problems, the Act prescribed two separate approaches.

First, there would be a two-step system of "technology based" effluent limits. Industrial sources would install the "best practicable technology" (BPT) by 1977, and then upgrade their treatment with the "best available technology" (BAT) by 1983.[11] The central feature of these controls is that sources must do the best they can afford to do.

Second, each state must develop a "water quality standard" for every water body within its boundaries.[12] These standards, known collectively as the "ambient approach," consist of a "designated water use" (for example, trout fishery, drinking water supply, industrial water supply) and a criterion for each appropriate pollutant set at sufficiently stringent levels to protect the use. State standards differ from BPT and BAT in that they focus on, rather than purposely ignore, ambient water quality conditions. They also differ from special toxic effluent standards because they incorporate a political (and therefore economic) judgment about desirable water uses rather than presume that all pollutant levels must be low enough to protect public health and the aquatic environment.

These three complementary approaches make up the universe of regulatory policy for water pollution control in America. They differ not only in the factors they emphasize (economics, ambient conditions, health threats), but most importantly in the level of scientific understanding needed to make each work. Much of the complex debate over toxic water pollution control can be understood in terms of policy choices between these theories.

THE FIRST POLICY SHIFT

By the mid-1970s it became clear that EPA had not fared well in its attempt to answer the questions posed by Congress. The Agency missed the Act's deadlines for listing toxic pollutants. When it did move ahead, EPA chose restricted criteria to select the chemicals, which focused on the prospect of discharges rather than on the chronic health hazards of pollutants. EPA began the rule-making process for six chemicals, and in 1975 became bogged down in disputes between environmental groups and industry experts. The Agency simply did not have the necessary scientific understanding to make Congress' control scheme effective.

In an effort to adjust EPA's priorities and accelerate the program, the Natural Resources Defense Council sued the EPA administrator four times between 1973 and 1975. These four cases were pending in U.S. District Court when, in late 1975, EPA approached NRDC with an offer to settle all four suits in a massive consent decree—to be signed and enforced by Judge Thomas Flannery before whom the cases had been consolidated.[13]

EPA offered to embark on a bold new program of regulating toxic discharges, which NRDC had the power to supervise through their ability to seek enforcement or contempt orders from Judge Flannery. After many months of negotiation, the parties agreed to a detailed settlement, and when the Judge approved it on June 8, 1976, the so-called Flannery Decree became the nation's blueprint for toxic water pollution control.

In practice, the Flannery Decree actually superseded the Clean Water Act. By settling its lawsuits, NRDC gave up its right to enforce the Act's regulatory scheme against EPA, and the Agency, in turn, placed itself under legal obligations to pursue an approach very different from what Congress had in mind.

The settlement addressed the question of "what is a toxic pollutant?" by actually listing 65 chemicals (now known as the "priority pollutants"). What Congress and EPA had been unable to decide, the parties simply negotiated. The list was crafted from an extensive survey of discharge-monitoring data and from numerous lists of known toxic chemicals prepared by various agencies. From a preliminary compilation totalling several hundred chemicals, the parties then bargained for a number they believed EPA could handle effectively. No further determination would be needed about whether these pollutants are harmful or, if so, in what amounts.

The Decree also determined who would be regulated and when. Twenty-one major industrial categories were selected (now known as the "primary industries"), and EPA agreed to promulgate final regulations for each category according to a phased schedule ending in December 1979.

The central feature of the Decree, however, was its prescription of a new method for regulating industrial sources. Thwarted by the rigorous scientific demands of developing toxic effluent standards, the parties decided to switch to the "technology based" approach utilizing BAT effluent limitations. EPA no longer would concern itself with toxic effects, but instead would examine the availability and costs of treatment technologies. Health and aquatic impacts would be set aside for the moment.

Additional provisions established a backup program for developing the next generation of toxic controls—a safety net in case BAT proved to be inadequate. EPA would derive numerical, ambient criteria for each of the priority pollutants. These could be placed in water quality standards as the states revised their rules in future years. In the near term, however, EPA would use these criteria as benchmarks to pursue a so-called hot spots program. The Agency would determine which water bodies still would not meet the criteria after BAT is implemented, and then develop site-specific strategies for going beyond BAT.

The Decree's new approach was attractive to all concerned. EPA was off the scientific hook, at least temporarily. The environmental groups would not obtain the stringent effluent standards prescribed by Congress, but they would see far more comprehensive regulations issued much sooner than if they had persisted in the litigation. Even the regulated industries were glad to substitute a one-time, complete set of regulatory requirements for the old "pollutant-of-the-month" approach that appeared to require a never-ending series of improvements as EPA added new chemicals to the list and developed toxic effluent standards.

Congress also thought the new approach made sense. Before reauthorizing and amending the Clean Water Act in December 1977, the House and Senate held a series of hearings to examine the Flannery Decree in detail. Congress strongly endorsed the BAT-toxics strategy, placing the priority pollutant list in the statute and confirming EPA's duty to issue regulations for the primary industries.[14]

Senator Edmund Muskie, the Senate's leader on clean water and chairman of the committee that drafted the 1977 amendments, captures the importance of the Flannery Decree:

> The Committee decided that the 1983 requirements based on best available technology . . . provide the most effective mechanism in the act for dealing with toxic pollutant discharges. Not only is BAT a technology-based requirement as recommended by the National Commission on Water Quality, but it also calls for heavy reliance on advanced technology and in-process controls which are appropriate for dealing with the majority of toxic discharges. More importantly, the nature of the toxics problem is so pervasive that the most effective approach in dealing with it is on an industry-by-industry basis.[15]

This strong endorsement led two federal courts to conclude that Congress intended for the court order to remain in effect and supply the details of the new strategy.[16] Thus, the Decree continued to limit EPA's discretion and to control the timing and scope of the nation's toxic pollution program.[17]

In his 1977 environmental message, President Carter directed EPA Administrator Douglas Costle to give the BAT-toxics program his "highest priority."[18] The Agency was trying hard, but struggling and failing to meet the Decree's deadlines. The parties had agreed to an ambitious schedule—leaving too little time to develop the analytical methods needed just to measure the priority pollutants in industrial waste streams. After NRDC sought a contempt citation against EPA, the parties again resolved their differences by agreeing to modify the Decree.[19] The Agency was given until 1981 to complete the program.

Despite these setbacks, EPA seemed to be developing effective regulations. The technical staff estimated that their rules would remove over 70 percent of the 400 million pounds of toxics discharged annually by industrial sources. By late 1980, 1 regulation had been issued and 11 more had been published in proposed form. The remainder were technically finished and could be issued close to the Agency's scheduled deadlines.

To date, this "technology forcing" never has received a true test. In theory, EPA establishes effluent limitations by pegging the allowable discharge level to that achieved by the best plants in an industrial category, and the rest of the industry is brought up to that standard. Unfortunately, in 1973–1974 when the first round of discharge permits were issued, EPA had not developed most of the necessary BPT effluent limitations. Over 90 percent of these permits were issued without the benefit of regulations; permit writers simply defined an individual BPT limit for each plant. The technology-forcing benefits of the categorical approach were lost because no categories actually existed.

The BAT-toxics strategy was to be different. For the most part, EPA and the states delayed issuing the next round of permits in 1978–1980 to await the Agency's effluent limitations. And the Decree provided that any new or reissued permit must contain a "re-opener clause," requiring these permits to be modified if subsequently issued regulations prescribed more stringent controls.[20]

Industrial sources have many options. They need not only add treatment equipment at the end of their discharge outfalls. Congress directed EPA to look for in-plant controls and changes in the production process as well. These might include ways to reduce the use of water in manufacturing, to substitute less-toxic materials, and to recycle or reclaim valuable chemicals usually lost in the wastewater discharge. "Best available technology" can include any of these alternatives. Some have proven effective, not only in curtailing toxic discharges, but reducing treatment costs as well.

In the petroleum refining industry, for example, the recycle and reuse of wastewaters can reduce discharge flows by nearly 40 percent. One-third of the plants in the industry already have this system in place; if the remainder adopted it, an additional 110,000 pounds of toxics would be removed annually without forcing any plants to close or affecting refining capacity at all.[21] Coke-making operations at steel mills could be using activated carbon and filtration technology to remove an additional 120,000 pounds of toxic metals and organics without any significant economic impacts.[22] Pulp mills can use the new Rapson-Reeve closed-cycle process to reduce their discharges by at least 75 percent. If the system is designed properly, it can achieve zero discharge, while saving energy, fiber, and water, as well as increase yield while decreasing chemical costs and treatment costs.[23] These are among a wide variety of control options available to EPA as it establishes BAT effluent limitations for the primary industries.

THE SECOND POLICY SHIFT—REVERSING A DECADE OF PROGRESS

When the Reagan administration took office early in 1981, the new administrator, Anne M. Burford, announced that EPA would no longer support nationally uniform, technology-forcing controls. BAT amounted to "treatment for the sake of treatment," EPA said, and was far less cost effective than ambient strategies might be.[24] The pendulum began to swing back toward the approach of the early 1970s that placed heavy emphasis on the comparisons of control costs with cleanup benefits and on the dominance of state and local preferences in the regulatory process.

Where the law calls for a three-stage process (BPT, BAT, hot spots controls), the new EPA now prefers two steps, from BPT directly to ambient, geographically limited regulations. The EPA leadership claims that, while BPT controls may have been aimed at conventional pollutants, they incidentally removed most of the toxics as well. Of 1.3 billion pounds of toxics that enter the raw waste streams of the primary industries each year, all but 412 million pounds seem to be removed by BPT-level technology—a reduction of 70 percent.[25] They also argue that the expense of further removal should be imposed only where one can document that it will make a worthwhile difference in ambient water quality.

The success of the EPA's new approach depends on three ingredients, none of which the Agency has the legal mandate or the technical skill to manage. First, EPA must now fight uphill against detailed statutes on issues that the courts and Congress heavily support. Second, the Agency's strategy cannot succeed without a thorough scientific understanding of toxic effects in the environment. This requires sophisticated aquatic modeling tools which can translate standards into effluent limits for numerous dischargers in hydrologically complex settings. It could take decades to collect sufficient data. Finally, this approach depends on enormous technical resources at the state and local level and on the ability of state enforcement authorities to use their often meager political clout against big industry, which usually constitutes a large portion of their tax base and plays a dominant force in the local employment markets.

Violating the Spirit of the Law

The Reagan administration wanted to move fast to change what President Carter's environmental leaders had set in motion. In one of their final actions, the Carter EPA officials had issued procedural rules in January 1981 to enforce BAT-level controls against industrial sources who discharge toxics into municipal sewers. Representatives of the chemical industry, automotive companies, and paper mills lobbied Vice President Bush to suspend the rules, and on April 2, 1981—three days after the regulation took effect—EPA did just that. Without the legally required notice-and-comment procedure, the new administration simply reversed what had taken years to put in place. One EPA official remarked in a press interview that "we were advised by Vice President Bush's Task Force on Regulatory Relief to delay the effective date, and procedural niceties were not involved."[26]

NRDC went to court again, and in July 1982 won a decision from the U.S. Court of Appeals in Philadelphia reversing EPA's move.[27] The regulations were restored, but over a year had been lost.

Next, EPA focused on the Flannery Decree itself. In April-May 1981, the steel industry and ten major chemical companies secretly lobbied Vice President Bush and EPA to join them in asking Judge Flannery to modify or vacate his order, calling the BAT-toxics program a "relic."[28] On June 4, 1981, EPA's general counsel advised Administrator Burford of industry's various options and recommended a strategy: based on alleged "resource constraints" caused by the Reagan budget cuts, she should ask the court to "stretch out" EPA's deadlines for most of the BAT rules and to eliminate several primary industries from the program.[29] Industry "would be pleased" with the recommended strategy, the memorandum concluded, but "obviously, the [plan] would not please the environmentalists."

Burford agreed to the strategy. On August 3, 1981, EPA asked the court to delay major portions of the toxics control effort and jettison the rest. Burford claimed that budget cuts had made the task more difficult and time consuming. But since EPA's staff had conceded that most of the work already was completed, NRDC asked to examine EPA budget documents that would reveal whether any such constraints were affecting the schedules. Rather than furnish the evidence, Burford asserted her now-familiar, "executive privilege" against disclosure.

Unable to plead poverty in a convincing manner, EPA changed course. In October, the Agency's inspector general began an investigation apparently designed to enable Burford to downplay her discredited claims that "resource constraints" were causing delays. The new EPA leadership planned to assert instead that since the Carter administration had left the program in such disarray, much more time was needed to straighten it out. When the inspector general issued his report, he concluded that "lack of resources was not indicated . . . as a significant reason for delays," and added that faulty management control and coordination were among the causes for not meeting the court's deadlines.[30]

Judge Flannery was not impressed. He rejected EPA's effort to undermine the program, not only ordering Mrs. Burford to issue the regulations promptly, but to find the necessary resources if they were lacking.[31]

Backed into a corner, EPA complied. Mrs. Burford shortened her program's schedules, found an additional $3 million to study and regulate the industries she had wanted to drop, and retained 14 staff positions she had planned to cut.

EPA is managing to issue the regulations according to its current court-ordered schedule. But a new, more subtle strategy emerges when one examines the rules themselves. Do they require the promising "best available technologies" such as those discussed above? Often the answer is "no." Many require absolutely no progress beyond BPT cleanup levels achieved by industrial sources in 1977.[32] Others require only marginal progress because they are based on end-of-pipe treatment systems rather than in-plant changes that can reduce specific constituents more effectively in a portion of the plant's discharge.[33] In all, the final regulations are considerably weaker than the proposed versions published several years ago by the Carter administration.

Much of the responsibility for these results apparently rests with officials of the Office of Management and Budget who hold a virtual veto power over EPA's actions. OMB has imposed an extra-statutory regime which effectively derails any regulation that it believes costs more dollars to remove a pound of pollutant than is prudent.[34] A promising technology will not be deemed "BAT" if it fails OMB's test—even though it may be the best available and industry clearly can afford it. The federal courts will have to decide whether this arbitrary cost-effectiveness test is valid under the Clean Water Act, as NRDC challenges regulation after regulation.[35] (See chapter 5 for a detailed account of the role of the OMB in recent EPA decisions.)

Meanwhile, what Judge Flannery refused to give, EPA now seeks through proposed amendments to the Clean Water Act. These would not only delay the court's deadlines, but also drastically weaken BAT controls. To date, EPA has requested the following amendments:

- Delay its deadlines to issue BAT regulations until 1984
- Delay industry's statutory deadline to comply with BAT from 1984 to 1988
- Give the administrator discretion not to issue new regulations for sources discharging into sewers and to rescind those already issued
- Allow municipal authorities to avoid enforcing the regulations against sources in their sewer systems
- Change the maximum duration of permits from five years to ten so that the Agency's weak BAT regulations will be untouchable for a decade.[36]

The silver bullet aimed at the toxics program, however, is the Chemical Manufacturers Association's proposal.[37] EPA supports this move in principle, but has not formally proposed it for fear of the inevitable backlash from House and Senate environmental leaders and the public.[38] The variance would transform a nationally uniform, technology based system into variable controls for each toxics discharger. An industrial source would obtain relief from BAT if it could show that it complies with ambient state water quality standards.

This variance proposal embodies the very concept rejected by the Flannery Decree and the 1977 Clean Water Act amendments. It was rejected then because EPA lacked the scientific understanding necessary to carry out an ambient-based approach. Has science advanced sufficiently in the last seven years to consider returning to the rejected system in lieu of firm, national BAT rules? Probably not.

The Role of Scientific Benchmarks

In order to secure safe controls not based on a technology approach, one must first establish valid *benchmarks* for determining how much of a chemical is toxic in the aquatic environment. The benchmark must be sound scientifically and, what is sometimes more important, acceptable legally to all concerned in the regulatory process. If any concerned party chooses to contest the scientific benchmark, EPA will sink back into the legal morass from which it emerged in 1976.

Under the Clean Water Act, the numerical criterion in a state water quality standard is the place to look for these benchmarks. State standards must be duly adopted through public procedures and then approved by EPA. They may be premised on the national criteria derived by EPA (pursuant to the Flannery Decree) or developed by state scientists using their own data and analyses.

Unfortunately, the states have not adopted criteria for the universe of toxic pollutants. Fewer than 20 states have any numerical standards for toxics. Those that do usually cover only a handful of heavy metals, and almost never consider chronic effects or exposures through the food chain. Only 16 states have quantitative standards for lead, cyanide, and cadmium; fewer than 15 have a PCB standard. Most of these standards are less stringent than EPA's national criteria, and the range of values (for any given pollutant) across the states is considerable. In EPA's recent regulatory impact analysis of a portion of the national toxics program, this sorry state of affairs led the investigators to conclude that "state water quality standards are too limited for determining the national water quality impacts of industrial [sources]."[39]

EPA's guidance is in no better shape. As required by the Flannery Decree, the Carter EPA issued national criteria in 1980 for most of the 65 priority pollutants. But Reagan EPA officials have repeatedly discredited the criteria, calling them scientifically flawed and unsuitable for states to use in setting their standards.[40]

Does this lack of benchmarks suggest a return to BAT? To EPA it does not; it dictates delay. In July 1981, EPA's head of research and development informed the administrator that the tools were not available for carrying out an ambient approach. He emphasized that:

- There are still no reference methods ... [for measurement of priority pollutants] applicable to marine waters, sediments or fish tissue.

- There are no low-cost practical screening techniques to scan large numbers of samples for the presence or absence of toxic pollutants in toxic amounts.
- Existing reference methods for toxic metals are not sufficiently sensitive.
- [There is an] inability to measure effluent discharge flow at the necessary level of precision to calculate receiving water impacts (and, therefore, waste treatment requirements to protect water quality).
- Existing models are inadequate for making wasteload allocations of the priority toxic pollutants.
- There are essentially no data available for assigning values to proposed incremental improvements in water quality, as would be essential in determining the highest level of water quality worthy of pursuit in a given stream reach.[41]

When will these tools be available? The EPA research and development chief said that "at the Fiscal Year 1981 level (and assuming zero inflation), the needed technical base could be completed, but not field validated, by approximately 1990."[42] EPA's research and development budget declined by 58 percent between 1981 and 1983 and by still larger amounts if one considers inflationary impacts on the Agency's purchasing power.

The other half of the safety net is in trouble as well. In addition to the criteria, each state standard must designate the water uses to be protected. In October 1982, EPA proposed a new regulation to govern how states establish these uses.[43] The proposal encourages downgrading from beneficial uses to less desirable ones, abandons protection of existing high-quality waters, and creates economic-hardship variances from toxics standards that are found nowhere in the statute. If these changes take effect, much less stringent criteria for toxics will be the norm.[44]

Limits to Local Initiative and Resources

The final test of EPA's alternative approach is the ability of states and cities to handle responsibilities formerly carried out by EPA. A regulator must have the resources and the clout to develop and impose complicated, expensive toxics controls on big industry. It might be premature to judge whether EPA's program passes this test, but there are disturbing signs that it will fail.

EPA's proposals call on the states and cities to derive ambient criteria for toxics, develop and enforce controls against sources who discharge into sewers, and administer variances of several kinds that require considerable expertise in both economics and health effects. State water pollution control programs, even without these new responsibilities, must be funded primarily by federal program grants. Tax revenues of state and local governments have never been sufficient to carry out Clean Water Act requirements without substantial assistance.

The Clean Water Act provides grants to each state to operate their water quality program.[45] States use these funds to develop standards, monitor water quality, issue permits, respond to chemical spills, manage other grant programs (for building municipal treatment systems), train municipal plant operators, and control runoff (non-point) sources of pollution. In fiscal year 1983, EPA gave the states over $54 million to meet these responsibilities.

In fiscal year 1984, the Reagan budget contains only $24 million—a 56-percent reduction at precisely the time states are being asked to assume more responsibility. In a February 1983 statement, the Association of State and Interstate Water Pollution Control Administrators had this to say about the President's cuts: "The President's budget . . . is a serious retreat by the Administration from the environmental goals of this Nation. It will devastate the state's ability to control toxic pollution. The timing is tragic given the expected increase in workload."[46] The fabric of the administration's alternative approach—state and local controls on toxics—appears to be shredding.

Finally, can the states and cities stand up to big industry? This concern was what led Congress to rely on national federal standards at the outset. If industrial sources threaten to locate in environmentally lenient areas, only uniform requirements can offset their ability to shop around.

In 1982, a particularly gruesome example occurred in Fort Wayne, Indiana. Because EPA had not issued national requirements for toxic discharges into sewers, the city was left to control its sources alone. In August, under pressure from local industries, the city council relaxed its requirements on cadmium, chromium, copper, cyanide, lead, phenol, and other highly toxic chemicals.[47] Industry representatives said the move would attract more industry and allow existing plants to expand. But many residents were concerned that increased toxic discharges would endanger people who unknowingly use contaminated municipal sludge for their vegetable gardens.[48] Dr. Edward Kimble of Purdue University's department of chemistry asked: "Is this the kind of industry we want for our city? Consider what sort of company would move to a location only because they can dump more chemicals into the Fort Wayne rivers than into the rivers near their home office."[49] The city fathers did consider Dr. Kimble's question, but their actions suggest that attracting any sort of company may be an acceptable trade-off for tough toxic chemical controls. *If* Fort Wayne is typical, the EPA strategy spells serious trouble.

America's toxic control effort, over a decade in the making, has been dismantled in just two years. Changes wrought through inaction, budget cutting, and regulatory "reform" cannot be completely reversed through the courts or by congressional oversight. It will take time to regain our footing—even if the federal administration changes abruptly in 1984—no matter which figure wins the day.

There are promising signs on the horizon. House and Senate leaders are on record against variances for toxics and have formally called on EPA to withdraw its proposal to weaken water quality standards.[50] Congress is considering amendments that would turn back some of EPA's regulatory reforms and restore the worst of the budget cuts.[51] And the federal courts may reverse several of EPA's weak BAT regulations.

There is hope yet for a comprehensive program to prohibit the discharge of toxic pollutants in toxic amounts. We may realize the national policy eventually, and one of the most important remaining questions is "When?" At the rate of over a million pounds of toxics discharged every day, we had better take that question seriously.

NOTES

1. Federal Water Pollution Control Act, as amended (Clean Water Act), sec. 101(a)(3), 33 U.S.C.A., sec. 1251(a)(3) (1983).
2. JRB Associates, Inc., *Assessments of the Impacts of Industrial Discharges on Publicly Owned Treatment Works, (Final Report to EPA)* (McLean, Va.: JRB Associates, Inc., 20 November 1981), pp. 1–3.
3 Lawrence Wright, "Troubled Waters," *New Times*, 13 May 1977, p. 34
4. Ibid., p. 35.
5. Ibid.
6. Ibid.
7. Ibid., pp. 35–36.
8. Ibid., p. 36.
9. *NRDC* v. *Train*, 8 ERC 2120 (D.D.C. 1976), as modified, *NRDC* v. *Costle*, 12 ERC 1833 (D.D.C. 1979).
10. Clean Water Act, sec. 307(a), 33 U.S.C.A., sec. 1317(a) (1983).
11. Ibid., sec. 301(b), 33 U.S.C.A., sec. 1313(b) (1983).
12. Ibid., sec. 303(a)(3), 33 U.S.C.A., sec. 1313(a)(3) (1983).
13. *NRDC* v. *Train*, 8 ERC 2120 (D.D.C. 1976).
14. *NRDC* v. *Costle*, 12 ERC 1833, 1836 (D.D.C. 1979).
15. U.S. Congress, Senate, Committee on Environment and Public Works, *A Legislative History of the Clean Water Act of 1977*, Committee Print 95–14, 95th Cong., 2d Sess., October 1978, pp. 454–55.
16. *NRDC* v. *Costle*, 12 ERC 1833, 1836 (D.D.C. 1979); *EDF* v. *Costle*, 636 F. 2d 1229, 1238-1245 (D.C. Cir. 1980).
17. *NRDC* v. *Train*, 8 ERC 2120 (D.D.C. 1976).
18. President James E. Carter, "1977 Environmental Message to the Congress," 23 May 1977.
19. *NRDC* v. *Costle*, 12 ERC 1833 (D.D.C. 1979).
20. *NRDC* v. *Train*, 8 ERC 2120, 2127 (¶10) (D.D.C. 1976).
21. U.S. Environmental Protection Agency, "Petroleum Refining Point Source Category Effluent Limitations Guidelines, Pretreatment Standards, and New Source Performance Standards: Final Rule," *Federal Register* 47, no. 201, 18 October 1982, 46438.
22. U.S. Environmental Protection Agency, "Iron and Steel Manufacturing Point Source Category Effluent Limitations Guidelines, Pretreatment Standards,

and New Source Performance Standards," *Federal Register* 47, no. 103, 27 May 1982, 23268.

23. U.S. Environmental Protection Agency, Office of Water and Waste Management, "Development Document for Proposed Effluent Limitations Guidelines, New Source Performance Standards, and Pretreatment Standards for the Pulp, Paper and Paperboard, and the Builders' Paper and Board Mills Point Source Categories," (Washington, D.C.: December 1980).

24. Anne M. Gorsuch, U.S. Environmental Protection Agency administrator, to the Honorable Thomas P. O'Neill, Jr., Speaker of the U.S. House of Representatives, letter to accompany EPA's 1982 legislative proposal, 25 May 1982.

25. JRB Associates, Inc., Table 1–1, pp. 1–3.

26. "EPA Sets Pretreatment Effective Date, Proposes Deferral Pending Impact Analysis," *BNA Environment Reporter*, 16 October 1981, 744.

27. *NRDC* v. *EPA*, 17 ERC 1721 (U.S. Court of Appeals—3d Circuit 1982).

28. Letter from American Iron and Steel Institute to Walter Barber, acting administrator of U.S. Environmental Protection Agency, 8 April 1981.

29. Memorandum from Office of General Counsel, U.S. Environmental Protection Agency to Anne M. Gorsuch, U.S. Environmental Protection Agency administrator, 4 June 1981.

30. Report from Matthew Novick, U.S. Environmental Protection Agency inspector general to Anne M. Gorsuch, U.S. Environmental Protection Agency administrator, 9 November 1981.

31. *NRDC* v. *Gorsuch*, 17 ERC 2013 (D.D.C. 1982).

32. Examples of regulations that require no progress beyond BPT include U.S. Environmental Protection Agency, "Ore Mining and Dressing Point Source Category Effluent Guidelines and New Source Performance Standards," *Federal Register* 47, no. 233, 3 December 1982, 54601; "Textile Mill Point Source Category Effluent Guidelines, Pretreatment Standards, and New Source Performance Standards," *Federal Register* 47, no. 171, 2 September 1982, 38814; "Leather Tanning and Finishing Industry Point Source Category; Effluent Limitations Guidelines, Pretreatment Standards, and New Source Performance Standards; Rule," *Federal Register* 47, no. 226, 23 November 1982, 52854; "Coal Mining Point Source Category; Effluent Limitations Guidelines for Existing Sources and Standards of Performance for New Sources," *Federal Register* 47, no. 198, 13 October 1982, 45385.

33. U.S. Environmental Protection Agency, "Organic Chemicals and Plastics and Synthetic Fibers Category Effluent Limitations Guidelines, Pretreatment Standards, and New Source Performance Standards; Proposed Regulation," *Federal Register* 47, no. 55, 21 March 1983, 11828–67.

34. Christopher DeMuth, U.S. Office of Management and Budget administrator for information and regulatory affairs to Joseph A. Cannon, U.S. Environmental Protection Agency associate administrator for policy and resource management (with attached OMB "Evaluation of Effluent Guideline Rules: Cost-Effectiveness as a Criterion"), 30 September 1982.

35. NRDC has filed petitions challenging, for example, the proposed regulations for the iron and steel point source category, *NRDC* v. *EPA*, no. 82–3464 (U.S. Court of Appeals, 3d Circuit, filed 8 September 1982), and the petroleum

refining category, *NRDC* v. *EPA*, no. 83–1122 (U.S. Court of Appeals, D.C. Circuit, filed 27 January 1983).

36. H.R. 6670 and S. 2652, 97th Cong., 2d Sess. (1982).

37. Chemical Manufacturers Association, "The Clean Water Act: A Background Paper," attachment to the "CMA Review of EPA's Proposed Amendments to the Clean Water Act," 5 April 1982.

38. James G. Watt, Chairman pro Tempore, Cabinet Council on Environment and Natural Resources, "Memorandum for the President: What Amendments to the Clean Water Act Should the Administration Propose?" 24 February 1983.

39. JRB Associates, Inc., pp. 2–4.

40. Frederic A. Eidsness, Jr., U.S. Environmental Protection Agency assistant administrator for water, "The Great Debate: Water Quality vs. Technology Based Standards," speech to the Water Pollution Control Federation, St. Louis, Mo., 6 October 1982.

41. U.S. Environmental Protection Agency, "Decision Unit Analysis," DU Code B 107, 11 September 1981.

42. Ibid.

43. U.S. Environmental Protection Agency, "Proposed Water Quality Standards and Public Meetings," *Federal Register* 47, no. 210, 29 October 1982, 49234–52.

44. U.S. Environmental Protection Agency, "Lists of Areas and Pollutants for Which Toxic Water Pollution Controls May Be Necessary to Protect Aquatic Life and Human Health in Compliance with S. 12 of the Consent Decree," 27 July 1981.

45. Clean Water Act, sec. 106, 33 U.S.C.A., sec. 1256 (1983).

46. Association of State and Interstate Water Pollution Control Administrators, "State Water Administrators Assess Impact of President's Budget," Washington, D.C., 1 February 1983.

47. Barbara Olenyik Morrow, "Council Asked to Ease Industry Clean-Water Rule," *Fort Wayne Journal-Gazette*, 25 August 1982, p. 1A.

48. Barbara Olenyik Morrow, "Gardeners Will Find Metals Mixed with Sewer Sludge," *Fort Wayne Journal-Gazette*, 26 September 1982, p. 1A.

49. Dr. Edward A. Kimble, "What Would We Attract?" *Fort Wayne Journal-Gazette*, 28 August 1982, p. 1.

50. U.S. Senators John H. Chafee, Robert T. Stafford, Jennings Randolph, and George J. Mitchell, and U.S. Representative Robert Roe to Dr. John W. Hernandez, deputy administrator of the U.S. Environmental Protection Agency, 14 March 1983.

51. S. 431, 98th Cong., 1st Sess. (1983).

5

The Politics of Toxic Wastes: Why the OMB Weakens EPA Programs

BRUCE PIASECKI

Many assume that America fails to rectify the toxic waste crisis because of poor leadership in the Environmental Protection Agency. Once this agency is purged of its bad blood, the public seems to believe, policies will improve and legislators can begin again their task of arresting toxic contamination. The same comfortable assumption colors most accounts of recent controversies over the Reagan EPA. The story often told is that ex-EPA head Anne Burford singlemindedly reneged on the agency's congressional mandate, delayed its court-ordered timetables, and openly defied public interests in a quest to placate private industrial requests.

Surprisingly, a reconstruction of the record[1] shows that this version is incomplete at best and often untrue. Anne Burford was fighting an uphill battle against demands from the Office of Management and Budget, whose mandate was to supply regulatory relief and to satisfy the President. The means by which the OMB influences EPA's programs tells us much about the making of toxic waste regulations in America. Here is how the process of OMB involvement in EPA rulemaking started.

THE LEARNING CURVE

In October of 1981, a group of EPA experts met to discuss dumping.[2] Their task was to improve toxic waste dumps, to strengthen them, so that they could contain the wastes without leaking. The problem these officials inherited—the appalling state of most U.S. landfills—had reached the public only a few months before. Ten thousand toxic dumps lay scattered across the U.S., with at least two thousand of them presenting serious threats to public health.

Gary N. Dietrich, then director of the EPA's Office of Solid Waste, led the October meetings. A thorough, exacting Cal Tech engineer, Dietrich had been with the EPA since its birth. He organized the first Office of Toxic Substances; solicited the first surveys that led to the creation of the Superfund, a national account for the emergency cleanup of dumps; and for several years, managed the EPA's entire budgeting process.

Earlier efforts to regulate dumps had been frustrated by incomplete industrial data.[3] So Dietrich's approach was a conceptual one based on engineering principles. He wanted to improve dumping at two stages of performance: first, during the active life of a landfill, the months to years when a dump is receiving wastes, and second, after it is full. The October meetings ended with the establishment of this double performance standard.

The second task proved easier than the first. By building an impermeable plastic cap on top of the dump, you can force rainwater to drain away from the dumps and cut the amount of moisture in the soil by 99 percent. Dietrich felt that this should be the principal line of defense. By limiting the amount of moisture, you cut considerably the generation of leachate—precipitation made toxic from percolating through wastes.

But the bigger problem remained control of wastes during a dump's active life. Rain or dew or disposal spills falling into the open pits creates "leachate clouds"—fine-particled pools of contaminants that follow an irreversible path into soil and groundwater.

Early in 1982, Dietrich announced his answer: an efficient liner placed at the bottom of the site would force leachate to flow into a collection drain before the final cap is put in place. The goal was to intercept leachate before it reached the soil and then pump it back out for storage or detoxification. Tests demonstrated that clay liners catch the runoff only some of the time.[4] So Dietrich opted for synthetic liners, sheets of plastic with the promise of greater resistance.

In February of 1982, Dietrich began circulating a diagram of a secured landfill. Looking like a bathtub with a tight-fitting umbrella overhead, the concept next passed to Dietrich's senior associates. EPA's top attorney, Robert Perry, liked the notion: it would relieve corporate fears of liability created by Love Canal. Although EPA's budgeting experts were nervous about increasing the cost of dumping, they knew immediate action was necessary. Within a week, Anne Burford approved the concept. It produced what she liked to call "environmental results." By the end of March, Dietrich's plans were ready for EPA's Red Border Review, a system for drafting rules that translate concepts into regulations ready for publication.

The goal of EPA's Red Border Reviews is to produce a regulation that can please presidential advisers yet can still direct industrial reform and appease environmentalists.

Dietrich's liner requirements had passed the Red Border Review when the EPA received a series of calls from the Office of Management and Budget. OMB understood the technical efficacy of a final cap; they even appreciated its political tact. But OMB's Office of Information and Regulatory Affairs expressed a strong dislike of liners. Dietrich noted that it wasn't that they felt liners cost too much, but that OMB thought liners might set an unacceptable precedent.

Dietrich had watched this cat-and-mouse game before. It is OMB's position that if you let an agency establish a precedent in an area otherwise devoid of regulation, then you might incite a series of costly new controls. In this case, OMB felt that if you demand liners for new landfills, then a court case could be won that required the thousands of America's old dumps to be fitted for liners. OMB sat on Dietrich's proposal for a few months, suggesting that the request for liners be deleted from the proposed regulation.

Meanwhile, Anne Burford was struggling to reachieve credibility from a previous decision. A month before, on February 25, she had reversed a ban that made it illegal to dump drums of toxic liquids into landfills.[5] The reasons for the ban were simple.[6] Liquids contaminate faster than solids; moreover, when they leak from storage, the drums that once contained them often crush under the weight of the overlying earth. This in turn breaks the seal of dirt meant to restrict water from entering the landfill. The water then mixes with the toxic chemicals, creating leachate clouds. Under this ban, thousands of companies had been forced to treat their wastes before dumping by dehydrating them into solids (which are easier to handle) or by detoxifying them via various chemical, physical, and biological processes.

Burford had reversed the rule, on a 90-day trial basis, as part of the President's program of regulatory relief. But the public outrage, heightened by complaints from industries that had invested in treating their wastes, forced the EPA to reinstitute the ban after 27 days.

The intensity of the public's concern over the dumping of liquids shocked many at the EPA. The EPA was not alone in its surprise. In response to this fiasco, the White House set up an alert memo system by which the EPA was required to forewarn White House officials about forthcoming decisions that might stir controversy.[7]

This alert system served as Burford's access to the White House. Unable to convince OMB reviewers of the EPA's needs, Burford carried EPA's proposed liner requirements up to the Vice President before her demands were met, notes Dietrich. On July 15, 1982, incited by a series of lawsuits that were about to damage Burford's image beyond repair, the White House directed the OMB to approve the EPA's liner regulations.

Although the net result is that all newly constructed dumps will install synthetic liners, the White House conceded to Burford's requests for reasons far different than Dietrich first imagined. Liner requirements were simply the most expedient way to create an impression of concern in a time when the White House was shocked by the public's wrath against dumping toxic liquids.

A final irony remained. Dietrich knew that liners were not enough. He had introduced the concept and fought hard to keep it alive, so as to direct the new administration toward a stronger posture on toxic dumping. Liners and precipitation caps were only the first two of a six-step plan to upgrade land disposal. But he knew, even back when he first sold Burford on liners, that they were not the complete answer.

"First of all," Dietrich explained, "it is very hard to install liners without holes or rips. Although one can hope that actual installation works most of the time, landfilling is a mining operation; it isn't easy to spread a thin sheet of plastic over 10 acres of dirt without rips."

Moreover, EPA's technical staff understood that liners wouldn't hold back many highly corrosive wastes. Although there are about two dozen different types of liners on the market, not one is resistant to all toxic wastes. So liners might not do much for dumps unless other regulations segregated wastes into dumps with appropriate liners. Dietrich also believed many wastes could be detoxified, with only their remnant parts landfilled, if proper regulations were passed.[8]

But Dietrich and Burford fought hard for liners, hoping to start widening the margin of safety. It was the only way they knew to make a philosophical showing to industry and the President. And it worked. In a period of extensive deregulation, the EPA was able to put forward the notion that landfillers should prevent leaks before they close a dump. Needing to grasp at something, they got the White House to admit that improving land disposal was a dire necessity.

The task of upgrading dumps, first assigned to the EPA in 1976, had become a six-year odyssey. Since the rest of his programs were likely to receive similar resistance, Dietrich left the EPA in May of 1982—a few months before Reagan's 11 top political appointees were purged from the agency in a series of congressional investigations that came to be known as "Sewergate."

THE ASCENT OF JIM J. TOZZI

The preceding tale by no means exhausts the range of forces influencing the EPA. But it does indicate a new tendency in rule making.[9] Congress, the courts, and the press have molded the shape of the EPA in highly visible ways, employing the pressures of public comment and the conventional checks of regulatory politics. Yet these approaches have been declining in relative importance. In their place, the Office of Management and Budget has evolved a system of checking the EPA that is far more discreet and that represents a movement in Washington that is strongly ascendent.[10]

It is difficult to say how much the current role OMB plays in rule making can be credited to a single parent, but it is beyond question that the growth of its power is reflected in large part by the arc of Jim J. Tozzi's career. As part of a small clan of career bureaucrats whose mission is to oversee federal agencies and coordinate their programs, Tozzi is one of Washington's premier regulatory managers.[11]

OMB's top 30 political appointees are housed in the Old Executive Building, an ornate, five-story structure built in England's "second empire style." These individuals, whose appointments average less than three years, work within a stone's throw of the President.

Around the corner and outside the black gates of the White House complex stands the New Executive Building—a tall, flat, red brick sky-scraper. Over 600 OMB federal employees are housed here, an icon to the mushrooming complexity of those who manage the executive branch, and with it, the fate of federal programs.

Tozzi came to the OMB in 1972, as part of its first wave of management experts. The old Bureau of Budget had consisted of civil servents preoccupied with pecuniary affairs. But when Nixon created the OMB, he added a number of appointments meant to assess broader questions of policy.[12] These new managers would not only check to see if each agency spent its money properly—they would also remind agencies of their goals. While most of these managers were restricted by the whims of their political appointments, Tozzi came in as part of the career staff housed in the New Executive Building.

Having developed the budget for the Corps of Engineers, served as a congressional liaison for the army, and assessed weapons systems for the Pentagon, Tozzi was oddly equipped for his first OMB job: monitoring the activities of Nixon's other new venture—the Environmental Protection Agency.

Tozzi remained chief of the OMB's Environment Branch for the next eight years, serving as the principal adviser on federal environment programs for four Presidents. His forte was programs involving poison. A graduate in chemical engineering from the Carnegie Institute of Technology, Tozzi came to be known as the final arbitrator. As changing administrations confronted problems with risk assessment, cancer policy, and leaking landfills, the name Jim J. Tozzi began to connote the position of the permanent government to a long string of EPA officials.

In the last hours of the Carter administration, Tozzi's brain child—the Paperwork Reduction Act of 1980—was signed.[13] Fearing that their computer operations might be restricted, the Defense Department and National Security Council initially opposed the bill; but by aligning business groups who felt the act might lighten government control, with key Carter advisers who sought an election year achievement, Tozzi resuscitated the bill after the November elections.

In signing the paperwork bill, Carter gave OMB a legal mandate to kill unnecessary government forms and eliminate burdensome items. Thus, Tozzi came to review the efficacy of all government paperwork—from the billion-dollar ventures of his old friends in the military establishment to the common 1040 tax form. Tozzi used this bill to create a single office in the law, whereby the President's men could check and coordinate the activities of each federal agency.

Soon after Reagan assumed office, Tozzi's program grew to consume the third floor of the New Executive Building. Seventy assistants came to staff the operation. Since every regulation of the government now had to be approved by Tozzi's floor, OMB secured a quiet power that made agencies respond more directly to its requests.

By the time the regulations were ready for Tozzi's review, they were often three or four years in the making. Sometimes, as in the case of Dietrich's landfill requirements, they had been ripening for half a decade.

Tozzi explains why these acres of proposed regulations matter: "Three fuels feed the federal engine: Dollars, People, and Rules. You can often keep control by cutting dollars or people. But sometimes a few people with a little funding can do a great deal of damage. In establishing the Office of Information and Regulatory Affairs, the OMB put its hand on a new, more effective throttle."[14]

Tozzi added, "But it wasn't until Reagan's Executive Order 12291 that we found the right tool. After four administrations, eleven years of contact with agencies, and fifty volumes of Quality-of-Life Reviews, we've sharpened an instrument that can now deliver regulatory relief."[15]

WHERE THE MUSIC IS MADE

Tozzi played a principal role in writing Executive Order 12291. Under this order, a proposed regulation is submitted to OMB at least 60 days *before* it appears in the *Federal Register*. At this time, the agency involved also submits a detailed evaluation of the regulation's anticipated economic impact. OMB then reviews both the RIA (Regulatory Impact Analysis) and the proposed regulation for consistency with the larger objectives of the President.

If questions arise during the review, OMB consults with the agency. At least 30 days prior to publication, the agency must submit the revised regulation along with its complete RIA. If OMB questions this final package, the agency is expected to refrain from publication until it satisfies OMB.[16]

You can think of 12291 in terms of the iron triangle, a common visualization of federal checks and balances. In one corner resides committee staff, congressional leaders with their own strict agenda. In another, the heads of each federal agency stand. The third corner contains special-interest groups, parties with power and money who want something.

Traditionally, once these groups start closing in on an issue, each participant doesn't have that much room to move. But 12291 gives OMB the lever to tilt the triangle on any of its sides so that they can slip in or out on behalf of the President. Although this central review process does not allow OMB to intervene at will, it does give OMB a chance to disrupt, if not change, the balance of power.[17]

Efforts to prevent toxic emissions into waterways are a case in point. While implementing programs that controlled direct dumping into surface waters, the EPA was sued in 1976 for failing to prevent indirect discharges—tons of toxic effluents which pass invisibly through conventional sewage treatment plants. Cornered by the court order, EPA spent the next three years identifying the ways industry should detoxify wastewaters before dumping them into sewage plants.[18]

After 4 public hearings, 16 public meetings, and over 400 comments from the regulated industries, the EPA announced their position. Instead of attempting to police thousands of indirect effluents across hundreds of industries, EPA would publish a series of timetables that stipulated when a particular kind of industry must install a specific pollution control device. It was called the BAT approach, effluent control by commanding the use of industry's "best available technologies."

On January 7, 1981, in the last days of the Carter administration, the EPA proposed its first BAT regulation: a complex, exhaustively detailed set of guidelines, which identified the dates and techniques by which the steel and iron industries would begin to control their toxic discharges.[19]

Five years in the making, this first BAT regulation was an embodiment of due process. But to Jim J. Tozzi, "Carter's steel guidelines had all the marks of excessive rule-making. It required the use of premature technologies, which weren't worth the cost. . . . And it would have sent most U.S. steel producers down the tubes."[20]

Throughout 1981, OMB and EPA compared notes on Carter's proposed rules. Since it was the first of a series of toxic effluent regulations, and one that could cost industry over a billion dollars to meet, Tozzi made the steel standards the first regulation to receive a complete 12291 Regulatory Impact Analysis. While Tozzi's staff now performs most of the steps of a review, he ran this package through himself.

On April 8, 1982, 15 months after Carter's initial proposal, the first clear test of 12291 was at hand. Acting as host, Tozzi had invited representatives from U.S. Steel, Bethlehem Steel, and the American Iron and Steel Institute. Also present were a representative from the Natural Resource Defense Council (NRDC), the environmental group that had successfully sued the EPA to enact BAT rule making and Joseph A.

Cannon, EPA's administrator for policy management. A series of memos from Cannon to Anne Burford serve to reconstruct the meeting.[21]

U.S. Steel claimed compliance for their company alone would cost over $255 million. EPA's analyses placed the cost around $80 million. Tozzi reminded the group that the Vice President, as well as OMB, was greatly concerned about the magnitude of this discrepancy.

The NRDC agent claimed U.S. Steel's estimates were dubious. When asked why U.S. Steel kept their year-old data until two weeks before the regulations were to appear in the *Federal Register*, Dr. Phil Masciantonio, representing U.S. Steel, said: "We did not feel it was germaine at the time."

Joseph Cannon claimed the problem was simple. In relying on a dated consultant's report, U.S. Steel's higher figures did not reflect the significant cost reductions EPA had enacted in its own Red Border Review, nor major cuts resulting from their year-long negotiations with OMB. Cannon also reminded Tozzi that the EPA needed to present a serious, steady face to the nation on toxic discharges.

For all its apparent potential as a grand showdown, the meeting ended quietly. Reminding the group that due process was almost overdue, Tozzi suggested that EPA and OMB resolve the problem themselves.

The next week, in an alert memo to Edward J. Rollins, assistant to the President for political affairs, Cannon reported: "We have reached an agreement with OMB." Carter's billion-dollar BAT program for steel had been cut to $310 million, a reduction of $724 million to industry in compliance costs. In addition, 21 older plants were to be given exemptions to the rule, on a case-by-case basis, should retrofitting prove too costly. Without once showing his cards, Tozzi had changed not only the direction of BAT rule making but also its basic arithmetic.

Tozzi learned a lesson from this steel review: "12291 let us clean up a regulation without messing in judiciary and congressional reviews." In terms of the iron triangle, OMB also learned that "for a traumatized rule maker like the EPA, OMB can serve as the lightning rod."[22] Cannon had warned the White House to expect major press coverage on the reduced rules, but none had occurred. EPA learned from this that if it passed the heat over to OMB, it would disappear.

The clock is not OMB's enemy; in fact, the OMB is most effective when the agencies under review are nearing deadlines. But OMB officials lost no time in using Tozzi's steel precedent to reflect on Reagan's larger mission. On May 11, 1982, Christopher DeMuth—OMB's spokesperson on regulatory affairs—sent a letter directly to Anne Burford. While Burford was still attempting to recover from the liquids controversy, OMB wanted her to understand that Tozzi's work on steel was only "the first important rule in a new round of rulemaking, covering twenty-six industries."[23]

Admitting that OMB didn't know "what exactly is the right amount to spend on removing a pound of water pollution," DeMuth asked the EPA to touch up on its economics: "I believe," the letter concluded, "that further work on developing cost-effectiveness data and cost-benefit benchmarks should be an important part of EPA's efforts."[24] The show had only begun.

No one quite knew yet what "cost-effective" rules looked like, yet the results kept coming in. EPA alert memos, sent to the White House following DeMuth's request, indicate significant cost reductions in control plans for battery manufacturing, paper, pulp, and paperboard industries, metals manufacturing, and the inorganic chemical industries.[25] Working in the shadow of the OMB, EPA excused a number of poisons from control, including hydrofluoric acid and hydrogen cyanide, before OMB review. In another case, OMB asked the EPA to revise pending controls on chloroform and benzene—highly toxic substances with ten-year records of contamination.[26]

Although the task of each OMB request differed, the effect upon the course of the EPA was the same. Traditionally, the EPA assessed the economic achievability of a regulation by computing four items: plant closures, employment impacts, investment requirements, and annual compliance costs. In addition, since the courts had ruled that economic analysis could not serve as the sole basis for BAT decisions, the EPA also surveyed pollution control hardware. To ensure that all industries could afford the selected technologies, EPA officials visited plants, counting the sites that already employed the endorsed devices.

After DeMuth's request, however, EPA's feasibility studies played a minor role. "You can think of cost-effective rule making as a new kind of federal traffic control," says Gary Dietrich. "For years, the EPA says the speed limit will be 55, so everyone travels 58. But if OMB says new numbers claim that you don't save more lives until you arrest folks going 70, then you'll find a new kind of cop on the beat."[27]

On February 23, 1983, EPA officials sent a letter to Tozzi, reporting the results of their crash course in economics. Three new ingredients have been added to rule making.[28] To assess whether industry is passing the costs to the consumer, EPA must inform OMB about changes in the price of the regulated products. To ensure that the new control does not stifle America's economic growth, EPA must provide information on changes in the availability of capital across the industry. To understand if the regulation might change the status of American goods in the world economy, EPA must also assess its international trade implications. By the time Anne Burford left, the EPA had become a budget shop, an agency more obsessed with economics than enforcement of controls.[29]

EDITING FEDERALISM

To this day, the extent to which OMB has authority to act as a surrogate President on highly technical matters like hazardous waste control, as well as the constitutionality of OMB's role in establishing cost-effective rule making in the agencies, is not clear.[30] Where OMB's power is successfully exerted, it rarely resembles that of conventional checks and balances. The normal access points by which Congress can cut funds or the courts can scrutinize presidential decisions are not available.[31]

Since OMB reorganizes and redirects an agency by modifying its decision models, it leaves only the faintest markings—scratches on the agency's final regulations. But the process by which OMB comes to its decisions and the day-to-day activities by which it revises proposed rules usually remain invisible. Thus, when Congress or the courts examine the final draft of a controversial toxic waste regulation, such as Dietrich's liner requirements, they find in it very little that directly reflects the OMB but instead evidence everywhere of the false starts, struggling, and internal modifications of the EPA.

It is only when an agency head protests hard enough, and remembers precisely the details of its original draft, that the role of OMB's editing becomes visible. A final instance reconstructs EPA's open resistance to OMB erasures and serves to illustrate a process of rule making that continues to grow long after Anne Burford's demise.

By court order, the EPA had to propose BAT controls for battery manufacturing by October 31, 1982. Lead, cadmium, and mercury, three of the most toxic metals in wastewaters, were the major pollutants of concern.[32]

On August 23, Tozzi alerted Christopher DeMuth to the EPA's proposed guidelines. On September 3, DeMuth told Tozzi's contact at the EPA, Joseph A. Cannon, that the OMB "found it necessary to extend the normal review period for this rule." This sent the EPA scrambling.[33]

By September 10, the EPA had secured its position paper. Three key individuals performed the exhaustive Red Border Review: Mahesh Podar, head of the Economic Analysis Division; Richard D. Morgenstern, director of the Office of Policy Analysis; and Joseph Cannon himself, associate administrator of the Office of Policy.[34]

These EPA reviewers verified that the new rules would affect only 165 plants, since 78 American battery facilities had already achieved zero discharges for all the pollutants under question. Although it would cost roughly $24.8 million to install these control technologies, and another $6.4 million in annual costs, the EPA did not expect these regulations to change either the rate of entry of new businesses into the industry or the growth rate of the industry, since price and profit changes would be less than one half of a percent. They also assured Burford there would be no plant closures, and that environmental groups were expected to support the plans since they would reduce discharge of toxic pollutants by about 99 percent.[35]

The average cost of the regulation was $21.28 per pound equivalent, a cost far below the norm set by other anticipated BAT guidelines. Nevertheless, OMB didn't like one item in the package: lead. Although lead has been considered highly toxic since the 1960s, it currently costs $258 per pound equivalent to control.

On September 29, EPA officials met with OMB to examine their differences. Joe Cannon explained EPA's cost estimate: they calculated pound equivalency by multiplying the number of pounds discharged from existing plants by a weighing factor. Frederic Eidsness, then EPA's director of the Office of Water, noted that this weighing factor reflected a long-standing consensus in the scientific community concerning the perceived toxicity of lead. Robert Perry, EPA's general counsel, reminded OMB that this kind of cost estimating had legal standing under BAT.[36]

But Christopher Demuth felt EPA's recommended technology for lead was not cost effective: $258 per pound equivalent was $58 dollars above OMB's trigger mechanism, a tolerance level alerting OMB reviewers to regulations that might violate presidential requests for regulatory relief. The next day, Cannon received a memo from Demuth: OMB had concluded its review; EPA's plans were not consistent with 12291.[37] The entire battery package must be reworked.

Anne Burford took the offensive. She personally reviewed all the numbers, requesting an update on the available technologies. Throughout the first week of October, EPA reconstructed what it could about OMB's past decisions. It appeared that OMB worked on a sliding scale.

A cost-effective number of $100 or less was regarded as *prima facie* evidence of reasonableness. A cost-effective number between $100 and $200 per pound equivalent required evidence of acceptable economic impact upon the regulated industries.[38] But in all cases, a cost-effective number above $200 was considered unreasonable. Checking this pattern with other agencies, Burford decided to retain the results of EPA's Red Border Reviews and run uphill against the OMB with the lead issue.

She was on solid ground. Of the 184 lead battery plants, 65 plants, or 35 percent of the total, had already achieved control of lead. Fifty-one of the zero-discharge plants achieved their control by flow reduction, the technique that OMB challenged as not cost effective. Installing flow reduction in the remaining plants would be quite cheap, relative to most pollution abatement programs, less than $23,000 a year for each plant.[39]

On October 6, three weeks before the court's due date, Joe Cannon wrote DeMuth one of the EPA's most diplomatic letters. Promising that the EPA would add a preamble that made "clear that we'd be willing to consider standards based on less stringent control technologies in the categories you identified," he explained that Burford needed desperately to get the package out.[40]

Next, Cannon sent an alert memo to Craig Fuller, the White House adviser on hazardous waste. Although Fuller had secured a reputation of standing monolithically behind the decisions and apprehensions of the OMB, Cannon asserted that EPA had to pass the rule. If the White House wanted Burford to remain in office much longer, they'd have to ask OMB to take the heat off once again.[41]

Later that month, Burford sent her own alert memo to the White House: "I have signed this rule today. OMB has reviewed the proposal and supports its issuance."[42]

With the same deliberate silence it had assumed upon accepting the White House request to let Dietrich's liners fly, the OMB withheld final comment on lead. It had reversed itself abruptly, and its rationale remained aloof. Had they revised their benchmarks? Had the cost-effective numbers changed? Had they now established an exception to the steel precedent? "Not everything here is number crunching," Tozzi explained. "Sometimes you need to concede a few yards so as not to fumble the football."[43]

Despite the uncertainty that surrounds many OMB reviews, one thing is clear. As in the popular West Indies' dance, during which participants must pass under a limbo bar as it drops successively lower and lower, you must be nimble and quick to pass under OMB's stick. Of the 340 EPA regulations reviewed by OMB in 1982, 262 were approved without change, 48 proved consistent with 12291 after changes, 11 were returned to the EPA for complete reconsideration, and 9 were withdrawn by the EPA altogether. (Eight others were left unresolved because the EPA continued to send them over improperly; and two were passed with an emergency status.)[44] Although a specific breakdown on hazardous waste regulations was not available, Tozzi explained: "When balancing the regulatory budget of the United States, you've got to keep the EPA in line. Otherwise, they will retail you to death."[45]

Yet OMB's ambitions have not stopped with the EPA. Employing a cost-effective sliding scale not unlike that first discovered in the lead case, OMB now reviews 30 regulations in a day, 7,500 annually. By the middle of Reagan's third year, the pattern is clear: 12291 has provided OMB with a strong counterforce to the traditional process of rule making[46]—in which major health, safety, and economic issues were resolved through thousands of discrete regulations. Comparing the last 23 months of the Carter administration with the first 23 months of the Reagan term, the number of pages in the *Federal Register* has declined by 31 percent and the number of final regulatory documents has declined by about 26 percent.[47]

To some veterans in Washington, the strength of 12291 for preempting regulations is a sign that OMB has gone too far. Senator Gary Hart sees it "as an emblem of Reagan's distrust of bipartisan politics, and an example of his disdain for legislative process."[48] John Dingell, chairman of the House Oversight and Investigations Committee, and a leader of congressional investigations into the demise of Anne Burford's EPA, sees it as evidence that "OMB will continue to clip the wings of the EPA, thwarting any earnest attempts to control toxic contamination."[49] Murray Comarow, executive director of the original Ash Commission assisting Nixon in his creation of both the EPA and the OMB, also laments these recent developments: "Over the last three years OMB has taken a needed function of the executive branch and distorted it to emasculate regulations."[50] Although regulatory relief may be appropriate in certain areas of government, it often spells disaster for efforts at controlling toxic wastes. For without big government's explicit involvement in establishing standards, it will be far more difficult to shift the country beyond dumping.

NOTES

1. Many sources provided memos in the effort to reconstruct these recent events, the most notable being the Natural Resources Defense Council (NRDC). Under the Freedom of Information Act, NRDC secured hundreds of documents indicating OMB involvement in EPA decision making. I am also indebted to the Fund for Investigative Journalism and the editors of *Atlantic Monthly* for their funding of my interviews with numerous Washington officials on this subject, including Congressmen Gary Hart, John Dingell, and James Florio.

2. EPA's working schedule has been reconstructed by Gary N. Dietrich, then director of programs to secure dumps. The details of this section have been recorded by the author from a series of interviews with Dietrich in April and May of 1983. Dietrich's account also has been recorded in depositions before the U.S. District Court for the District of Columbia. For an example, see "Deposition of Gary N. Dietrich," 28 January 1982, Civil Action Suit No. 78–1899, published by Ace Federal Reporters, Inc., Washington, D.C.

3. For Dietrich's own account of the problems of insufficient data, see chapter 2.

4. For a more detailed account of the limitations of securing landfills, see chapter 1.

5. Khristine Hall reports on the role of private watchdog agencies in reinstating this liquids ban in chapter 3. Hall played a key role in the Environmental Defense Fund's (EDF) depositions of EPA officials during this time.

6. For an account of the role of industrial lobbying groups in reinstating the liquids ban, see B. Piasecki's "Struggling to Be Born: A New Industry Takes on Toxic Wastes and the EPA," *Amicus Journal*, Spring 1983 (vol. 4, no. 4), pp. 9–11. *Amicus* is a publication of the Natural Resources Defense Council.

7. Interview with Randy Mott, a Washington-based attorney then representing the Hazardous Waste Treatment Council (HWTC). HWTC was the industrial coalition fighting to reinstate the ban. See chapter 11 and note 6 for more information on HWTC.

8. Gary N. Dietrich, letter to *Science 83* magazine, May 18, 1983. (2 pages.)

9. For a detailed assessment of the possible unconstitutionality of OMB rule making, see Morton Rosenberg's "Presidential Control of Agency Rulemaking: An Analysis of Constitutional Issues That May Be Raised by Executive Order

12291," published by the Congressional Research Service of the Library of Congress on June 15, 1981. (145 pages.) Mr. Rosenberg is a specialist in American public law; he wrote this report in response to a request from the House Oversight and Investigations Committee, chaired by Congressman John Dingell.

10. Testimony of John D. Dingell, before the House Subcommittee on Legislation and National Security, U.S. House of Representatives, April 27, 1983. (8 pages.) According to Dingell, as stated in the above testimony, "It is only when we appreciate the full range of tools which OMB possesses and exercises to control an Agency's regulatory policies that we realize the necessity to clip some wings so that statutory mandates are observed" (p. 1).

11. For a synopsis of Tozzi's influential career, see Peter Behr's "If There's a New Rule, Jim Tozzi Has Read It," in the federal report section of the *Washington Post*, January 23, 1982, p. A21.

12. For those interested in the means by which Nixon centralized policy dictates and agency rule making in the OMB, see a series of recently declassified *Memoranda for the President of the United States*, submitted by the President's Advisory Council on Executive Organization (Roy L. Ash, chair), April 1969 to November 1970. Two items are of particular interest concerning the OMB-EPA connection. See "Federal Organization for Environmental Protection," April 29, 1979, pp. 121–42; and "The Executive Office of the President—An Overview," October 26, 1970, pp. 275–312. I wish to thank Murray Comarow, executive director of this council from July 1969 to July 1970, for his aid in gaining access to these documents.

13. For the best summary of the efficacy of the Paperwork Act, see "Managing Federal Information Resources: Second Annual Report under the Paperwork Reduction Act of 1980," published by OMB in April of 1983 under Section 3514 of the Act. (31 pages, with appendices and cover letters by OMB Director David A. Stockman to George Bush and Speaker of the House Thomas P. O'Neill.)

14. Interview with Jim J. Tozzi, on assignment from the *Atlantic Monthly*, May 4, 1983.

15. Interview with Jim J. Tozzi, on assignment from the *Atlantic Monthly*, May 5, 1983.

16. For a detailed assessment of 12291 and for an explanation of its procedures and requirements, see "Executive Order 12291 on Federal Regulation: Progress during 1982," published by the Office of Management and Budget in April 1983 and released to select federal officials on April 25, 1983. (31 pages, with charts.) For further information on this study, contact OMB's publications office at 202–395–3080. I thank Jim J. Tozzi for supplying copies of this and numerous other OMB reports.

17. I borrow the image of the iron triangle, and the description of how 12291 allows OMB to tilt the triangle, directly from Tozzi's own account (interviews with Jim J. Tozzi, May 4–5, 1983). For a detailed assessment of the possible unconstitutionality of OMB intervention, see the reference cited in note 9 of this chapter.

18. For an elaborately detailed account of efforts to establish reliable water controls, see chapter 4 by NRDC's senior attorney, James Banks.

19. Environmental Protection Agency, "Iron and Steel Manufacturing Point Source Category Effluent Limitations Guidelines, Pretreatment Standards, and

New Source Performance Standards," *Federal Register*, 47, no. 103, May 27, 1982, p. 23268.

20. Interview with Jim J. Tozzi, on assignment from the *Atlantic Monthly*, May 5, 1983.

21. These documents include the following EPA memoranda issued from the Office of Policy and Resource Management: "Final Rule Establishing Effluent Guidelines for the Iron and Steel Industry," sent to Anne Burford, EPA administrator; "EPA Action Item on Iron and Steel Effluent Guidelines," sent to Edward J. Rollins, assistant to the President for political affairs at the White House; and another "EPA Action Item" on the same subject sent to Craig L. Fuller, assistant to the President for cabinet affairs. All three of the cited memos were sent by Joseph A. Cannon, acting administrator for policy and resource management. Each alert memo was designated "Major Significance/Controversial."

22. Interview with Jim J. Tozzi, May 5, 1983.

23. Christopher DeMuth, administrator for information and regulatory affairs, a letter to Honorable Anne Gorsuch, May 11, 1983. (2 pages.) Copies sent to Joe Cannon and Steven Schatzow, director of EPA's Office of Water Regulations and Standards, with a chart on "Estimated Removal Costs for Source Subcategories in the Iron and Steel Industry."

24. Ibid., p. 2.

25. These memos were secured by NRDC under the Freedom of Information Act in the winter of 1983.

26. Interview with Francis Dubrowski, an attorney for NRDC, April 4, 1983. Legal documentation on these revised rules include: "Proposed Water Quality Standards and Public Meetings," *Federal Register*, 47, no. 210, October 29, 1982; a petition for review of these EPA actions by NRDC (no. 81–2068 before the U.S. Court of Appeals for the 3d Circuit), filed by Francis Dubrowski; NRDC's prepared "Comments on EPA's Proposed Revisions to Water Quality Standards Regulations" (submitted to the court on February 10, 1983); and a series of internal EPA memos following DeMuth's letter.

27. Interview with Gary Dietrich, on assignment for the *Atlantic Monthly*, May 4, 1983.

28. Richard Morgenstern, director of the Office of Policy Analysis and Steven Schatzow, director of the Office of Water Regulations and Standards, letter to James Tozzi, February 23, 1983. (2 pages.)

29. Joe Cannon, letter to Christopher DeMuth, October 6, 1982. (2 pages.) This letter also serves to document this major change in EPA's methods of economic analysis.

30. Interview with Congressman John D. Dingell, May 5, 1983.

31. Interview with Congressman Gary Hart, May 5, 1983.

32. "Summary of Issues: Battery Manufacturing Effluent Guidelines," NRDC document, n.d. I thank James Banks and his assistant Justin Ward for the documents sent to reconstruct this battery case.

33. Christopher DeMuth, letter to Mr. Joseph A. Cannon, September 3, 1982.

34. EPA memorandum, "Proposed Effluent Guidelines for Battery Manufacturers—Red Border," September 10, 1982. (3 pages, with attachments.)

35. Joseph Cannon, EPA memorandum to Anne Gorsuch, "Rule Proposing Effluent Guidelines for the Battery Industry." (2 pages.)

36. This meeting is recorded in a number of EPA memoranda; the most extensive account occurs in a memo to Anne Gorsuch titled "Cost-Effectiveness Analysis for Effluent Guidelines." (5 pages.) Copies of this exacting summary were sent to four EPA officials, including Joe Cannon and EPA general counsel, Robert Perry.

37. This OMB rejection is noted in a number of EPA documents, including the memo referred to in the preceding note.

38. OMB's sliding scale is reconstructed from a series of internal OMB and EPA draft documents leaked to NRDC.

39. "Cost-Effectiveness Analysis for Effluent Guidelines," pp. 2–3.

40. Draft of Cannon letter to DeMuth, October 6, 1982.

41. Joseph Cannon, EPA action item, to Craig Fuller, "Effluent Guidelines for Battery Manufacturers."

42. Anne Gorsuch, EPA action item, "Effluent Guidelines for Battery Manufacturers."

43. Telephone interview with Jim J. Tozzi, June 17, 1983.

44. OMB report, "Executive Order 12291 on Federal Regulation," p. 10.

45. Interview with Jim J. Tozzi, May 4, 1983.

46. "Executive Order 12291 of February 17, 1981," *Federal Register*, 46, no. 33, February 17, 1981, pp. 13193–98.

47. OMB report, "Executive Order 12291 on Federal Regulation," p. 4. Also, see "Materials on President Reagan's Program for Regulatory Relief," June 13, 1981. (108 pages.)

48. Interview with Senator Gary Hart, May 5, 1983.

49. Interview with Representative John Dingell, May 5, 1983.

50. Interview with Murray Comarow, May 6, 1983.

Part II
STATE AND EUROPEAN INITIATIVES FOR STRONGER CONTROLS

6
Discouraging Dumping: The California Example

GARY A. DAVIS

California has taken a bold step recently to reduce the public health and environmental impacts associated with hazardous waste management. In doing so, it serves as a model for other states and the federal government. On January 25, 1983, state regulations went into effect that will phase out the land disposal of five categories of hazardous wastes. The intent of these regulations and accompanying incentives and disincentives is to spur the creation of a system of recycling, treatment, and thermal destruction facilities for managing those wastes that present the greatest long-term risks when disposed of on or into the land.

The California program is based on two findings:

- Land disposal of many hazardous wastes, particularly those that are highly toxic, persistent, mobile, and bioaccumulative, poses unacceptable long-term risks to public health and the environment.
- Technologies do exist and are commercially available for recycling or detoxifying the wastes that present the greatest hazards when disposed of into the land. At present, they are not being used to any meaningful extent because of the availability of inexpensive land disposal.

Although these findings are not new or startling, California's program to restrict land disposal and to encourage the use of alternative technologies represents their first translation into concrete public policy in America. These findings, and the wisdom of a shift away from the land disposal of hazardous wastes, have recently been confirmed in a major study by the congressional Office of Technology Assessment.[1] Other states and the U.S. Congress are now considering similar programs.

THE CALIFORNIA PROGRAM

California has begun its program to reduce dependency on land disposal for hazardous waste management without creating an undue burden on industry in the state. The approach taken recognizes that in California, with areas of arid climate and little groundwater, land disposal of some hazardous wastes may be appropriate. It also prompts industries to take care of their own wastes by encouraging the development of legal and safe waste management options.

There are four basic ways for government to intervene in the market-place to correct practices having an undesired impact upon society:

- Government can undertake the activity itself to ensure that it is conducted properly
- It can mandate or prohibit certain practices by legislation and regulation
- It can provide positive incentives to encourage better practices
- It can provide disincentives to discourage bad practices

Many European countries have emphasized the first approach in promoting the utilization of alternatives to land disposal. In Denmark, for instance, a consortium of municipalities financed the construction of a centralized treatment facility, and each municipality provides a collection station for hazardous wastes. Some states in this country, such as New York and Arizona, are also leaning heavily on the first approach as they begin to create hazardous waste facilities that will emphasize recycling and treatment rather than disposal. Both states have plans to sponsor facilities that will be built on state-owned land and will be constructed and operated by contractors chosen by the state. California has chosen to rely on a combination of the other three approaches rather than to accept the potential liability inherent in the first.

The Regulatory Stick

The centerpiece of California's program is the regulatory stick, the set of regulations phasing out the land disposal of the following five categories of hazardous wastes by the dates shown below:[2]

- Cyanide wastes (June 1, 1983)
- Toxic metal wastes (January 1, 1984)
- Strong acids (January 1, 1984)
- PCB liquids (January 1, 1984)
- Halogenated organics (liquids—January 1, 1985; solids and sludges—July 1, 1985)

Other hazardous wastes will be added to the list in an ongoing process of identifying those hazardous wastes that are least appropriate for land disposal. These regulations go well beyond EPA restrictions on the landfilling of containerized and bulk liquids.

Concentration thresholds have been set for free cyanides, for listed toxic metals, PCBs, and halogenated organics. Acid wastes will be banned if their pH is less than or equal to 2.0. Land disposal methods prohibited by these regulations include landfilling, ponding, land farming, and deep-well injection, whether on site or off site. In addition, storage of the specified wastes in the land for longer than one year, or land treatment where the residues are hazardous and remain in the land for longer than one year, are also prohibited.

The dates have been chosen as projections of when facilities employing alternative technologies can be permitted and constructed in the state. These restriction dates give hazardous waste recycling and treatment firms a definite target as to when a market for their services will be created, but no waste will be banned before the recycling or treatment facilities are available to manage it. The regulations contain procedures for determining whether treatment capacity can meet the waste load demands before the restrictions take effect. This allows for input from hazardous waste generators, waste management firms, and the public.

The regulations also contain two variance procedures. One permits waste generators or disposers to demonstrate to the Department of Health Services that, even though their particular waste or disposal method would be prohibited by the regulations, the land disposal of their waste or use of their disposal method would not present risks to the public health or the environment. The other variance procedure allows the Department to grant emergency variances for the land disposal of spill materials and also to permit temporary land disposal if recycling and treatment facilities experience unexpected shutdowns and there is no other facility available to recycle, treat, or store the wastes.

Incentives to Encourage Alternatives to Land Disposal

The second component of the program is a group of incentives to encourage the construction and use of recycling, treatment, and thermal destruction facilities. The first of these is a financial incentive. The California Pollution Control Financing Authority is authorized to provide low-interest loans through the sale of tax-exempt state bonds for financing hazardous waste recycling and treatment facilities. The state also has expanded the California Waste Exchange, a successful waste clearinghouse operated by the Department of Health Services, to actively assist generators of hazardous wastes in identifying recycling opportunities. Under the statute authorizing its operation, the Waste Exchange can review hazardous waste manifests that accompany each shipment to find hazardous wastes that have the potential for recycling and require the generator of those wastes to justify why they were not recycled.[3]

Finally, as an incentive to the construction of alternative facilities, the Governor's Office of Planning and Research assists project proponents by coordinating the permit requirements of the several state and local agencies. This coordination can cut drastically the time required to obtain permits and can reduce the uncertainty in dealing with several different agencies. California officials expect to streamline the process further by consolidating some of the authority now vested in diverse agencies.

Disincentives for Land Disposal

The most important disincentives against continued dumping are two fees levied on the land disposal of hazardous wastes. These fees help narrow the cost differential that exists between the current price of land disposal and the costs of most treatment methods. The first fee, which was $4 per ton with a $10,000 per month cap in fiscal year 1982–1983, goes to support the Department of Health Service's regulatory program.[4] This fee will generate about $5.3 million in 1983, which combined with money granted to California under RCRA, allows the regulatory program to operate without general taxpayers' money. Proposed regulations will increase this fee to $11.45 per ton for those hazardous wastes that will be eventually restricted from land disposal and to $4.05 for all others. The second fee, which has different levels for different wastes according to a rough degree of hazard classification, is set to guarantee that $10 million per year is generated to be used for cleaning up uncontrolled hazardous waste sites in the state, for providing emergency response capability for chemical releases, and for compensating victims of chemical releases.[5]

DEVELOPMENT OF THE PROGRAM

Background

California was a leader in the development of regulatory programs to ensure safe management of hazardous wastes. The state developed the first "cradle-to-grave" hazardous waste tracking system in 1972;[6] this became the model for the federal Resource Conservation and Recovery Act.[7] The state also developed early criteria for the siting, design, and construction of land disposal facilities.[8] Although these regulations may have prevented widespread haphazard dumping of hazardous wastes in the state, officials began to discover in the early 1970s that all was not well with some of the state's land disposal facilities.

A vivid example of the failure of conventional landfill facilities is the Stringfellow acid pits in Riverside county, now a priority site on the federal Superfund cleanup list. The site was constructed in a box canyon in 1955–1956 as a series of evaporation ponds to concentrate liquid wastes.

It received approximately 32 million gallons of acids and caustics containing toxic metals and wastes containing DDT and DDE. It was closed in 1972, when it became apparent that wastes were leaking through and under containment dams. In 1978, during heavy rains, the state was forced to release 80,000 gallons of the wastes into the creeks and storm drains of nearby Glen Avon. This was done to prevent the containment structures from being breached altogether.

Other licensed sites were plagued with problems throughout the 1970s, and in 1980 three sites were closed following geological investigations indicating that they were not secure. When another site was closed after reaching capacity in 1980, the state was left with six operating Class I land disposal facilities (those that can accept any and all types of hazardous wastes) and only one within 200 miles of the Los Angeles basin where the majority of California's hazardous wastes are generated.

In response to this dramatic reduction in land disposal capacity, a number of hazardous waste generators pressured the state for new landfill sites. But the story of Love Canal had exploded in 1980, and a growing concern throughout the country about dumping caused Governor Edmund G. Brown, Jr., to resist this pressure. He called for a study of the problems of land disposal and the feasibility of establishing alternative technologies in the state. The goal was to obviate the need for new landfills.

Basis of the Policy Shift: The Office of Appropriate Technology Study

The key to California's shift from dumping was the study prepared by the Governor's Office of Appropriate Technology (OAT), *Alternatives to the Land Disposal of Hazardous Wastes: An Assessment for California.*[9] This study documented in detail the amounts and types of hazardous wastes that were being disposed of on or into the land and assessed the technologies that were available to recycle, treat, and destroy the wastes that are least appropriate for land disposal. The Office of Appropriate Technology was established by Governor Brown in 1976 to assist and advise the Governor and state agencies in developing less costly, less energy-intensive, and less environmentally degrading technologies relating to waste disposal, transportation, agriculture, and building design. Robert Judd, then director of OAT, assembled the Toxic Waste Assessment Group to be an interdisciplinary team, including a project manager with extensive state agency experience, a chemical engineer/attorney, a mechanical engineer with incineration expertise, and a resource economist.

Waste Characterization

The core of the study was a detailed characterization of waste volumes and types, performed by the University of California's Department of Chemical Engineering at Davis. Professor David Ollis, who had completed similar work at Princeton University for the Delaware River Basin Commission, directed the waste characterization. His team used the hazardous waste manifest forms that accompany every shipment of hazardous wastes in the state. They produced a "snapshot" of waste generation and disposal by documenting the amount of hazardous wastes generated and disposed of according to 95 different categories. They also identified the amount and type of wastes generated by each industry on a regional basis, and documented the amount and types of wastes disposed of by each land disposal technique and at each land disposal facility in the state.[10]

Professor Ollis' team found that there were about 1.3 million tons per year of hazardous wastes being sent to off-site commercial hazardous waste facilities in the state. Ninety percent of this was disposed of on or into the land. The major generators of hazardous wastes in California were the petroleum extraction and refining industries and the chemical industry, particularly agricultural chemical production. The major hazardous waste-producing regions were the greater Los Angeles/Long Beach area, the San Francisco Bay area, and the Bakersfield area.

Alternative Technologies for High Priority Wastes

In order to focus the investigation of alternatives to land disposal, the OAT team established general criteria for hazardous wastes that are inappropriate for land disposal and selected generic categories of hazardous wastes meeting these criteria. These were called "high priority wastes." The criteria were:

- High toxicity
- Persistence in the environment
- Ability to bioaccumulate
- Mobility in a land disposal environment

The categories of "high priority wastes" addressed in the report were:

- Pesticide wastes
- Polychlorinated biphenyls (PCBs)
- Cyanide wastes
- Toxic metal wastes
- Halogenated organics
- Non-halogenated volatile organics

Approximately 40 percent (about 500,000 tons per year) of those hazardous wastes sent off site for disposal fell into these six categories of high priority wastes.

By reviewing current technical literature, the Toxic Waste Assessment Group found that recycling, treatment, or thermal destruction technologies were commercially available for all the types of hazardous wastes in each of the subcategories. In addition, in a less detailed analysis, alternatives were found to be available for another 35 percent of the waste stream. This led to the finding that approximately 75 percent of California's off-site hazardous waste stream could be diverted from land disposal.

In reviewing the wide array of alternatives, the group established a priority list stipulating a sequence of preferred management strategies. It read as follows:

- Waste reduction: reducing or eliminating wastes at the source by changing industrial processes so that fewer hazardous wastes are produced or by switching to products that result in fewer hazardous wastes.
- Recycling: reusing hazardous wastes or recovering valuable materials from them.
- Physical, chemical, or biological treatment: many treatment processes have been developed that render hazardous wastes completely innocuous, reduce toxicity, or substantially reduce the volume of material requiring disposal.
- Thermal destruction: most organic hazardous wastes that cannot be effectively recycled or treated can be incinerated at high temperatures, or otherwise thermally decomposed, to produce relatively innocuous residues. Energy can also be recovered.
- Solidification/stabilization of residuals before land disposal: some hazardous wastes can be solidified into a rock-like form that reduces solubility and provides additional protection to groundwater.

The most widely used alternative technologies for hazardous waste management include:

Physical treatment
- Sedimentation
- Filtration
- Solar evaporation
- Distillation
- Flotation
- Absorption

Chemical treatment
- Neutralization
- Precipitation

Biological treatment
- Waste stabilization ponds
- Activated sludge

Thermal destruction
- Liquid injection
- Rotary kilns

There are also several emerging treatment and thermal destruction technologies that show promise in increasing the effectiveness of treatment or destruction and in reducing the costs and environmental pollution associated with these technologies. These were described in the OAT report and several are being demonstrated in California under a grant from the U.S. EPA. (Some of these are described in the concluding chapter; others are listed in appendix A.)

The Economics of Treatment vs. Disposal

The report also compared the costs of using alternative technologies to the continued use of land disposal for high priority wastes. The analysis found that it would cost approximately $50 million per year to recycle, treat, and destroy all 500,000 tons per year of California's high priority wastes as compared to approximately $20 million per year for land disposal. However, when these additional costs were compared to the total costs of doing business for California's industries, it was found that most industries would still have hazardous waste management costs below 1 percent of their total business costs. The additional $30 million per year for hazardous waste management would also be spread among industries having combined gross sales in excess of $30 billion per year.

To perform this analysis a recycling, treatment, or destruction technology was selected for each subcategory of high priority waste, and an average cost per ton of that technology was found in the literature or by contacting hazardous waste treatment firms. The total cost of using alternative technologies for all the high priority wastes was then compared to the current costs of using land disposal. Transportation costs were assumed to be equal with both scenarios even though recycling and treatment facilities are usually located nearer to waste generators than land disposal facilities. An analysis was also done to see which industries would be hardest hit by a switch to alternative technologies.

The economic analysis also took into account the potential for the land disposal facilities currently being used to become sources of environmental contamination. The costs of cleaning up these facilities would be astronomical, far in excess of the costs of recycling, treatment, or destruction of the hazardous wastes. The California Department of Health Services has estimated that it will cost in excess of $300 million to clean up 67 of the known uncontrolled sites in the state. This led to one of the report's major findings: although land disposal may be cheaper in the short run, in the long run it is a misleading bargain.

Recommendations of the Report

The OAT report concluded with the recommendation that the state abandon the concept of Class I landfills as repositories for any and all types of hazardous wastes and, instead, begin a program to reduce dependence on land disposal and encourage the construction of facilities for recycling, treatment, and destruction. Specific recommendations were those that shaped the program described above. The recommendations of the report were accepted by Governor Brown and incorporated into an executive order directing the Department of Health Services to use existing statutory authority to implement them.[11]

Development of the Land Disposal Restriction Regulations

Developing the regulations to phase out the land disposal of high priority wastes was a painstaking and often controversial process that took over 15 months. The publicity surrounding the Governor's executive order was somewhat misleading and engendered an initial outcry of opposition from industry in the state. Hazardous waste generators felt that the restrictions were unnecessary and would leave them without legal disposal methods. Even the hazardous waste management industry in the state, which stood to gain from a program that encouraged alternatives to land disposal, was apprehensive. Some of this apprehension was due to the fact that most of these companies also operated land disposal facilities and depended on the goodwill of hazardous waste generators for their livelihood.

An Interagency Task Force was created to develop the regulations to phase out the land disposal of high priority wastes. The task force consisted of representatives from the Governor's Office of Appropriate Technology, the Department of Health Services (the agency with general responsibility for regulating hazardous wastes), the State Water Resources Control Board (responsible for regulating discharges to groundwater from land disposal facilities), and the California Air Resources Board (responsible for controlling air emissions). In addition, task force members assisted in the coordination of permit requirements of their agencies to expedite permission for alternative facilities.

The controversy generated by the executive order led the California Assembly Committee on Consumer Protection and Toxic Substances to call for a hearing on the merits of the program, which was held in February 1982. For eight hours, experts from all over the country debated the necessity of restrictions on land disposal and the proposed manner of imposing them. The Chemical Manufacturers Association, the Western Oil and Gas Association, and the industry's lobbying organization, California Committee on Environmental and Economic Balance, were the most vocal critics of the program. Supporters included a New York State hazardous waste official, hazardous waste management firms, local government officials, and several environmental groups. A representative of SCA Services, a hazardous waste management firm operating several treatment facilities in other states, disputed the claim by hazardous waste generators that the switch to alternatives would be too expensive, telling the committee that the OAT study had actually overestimated the costs of recycling and treatment. At the conclusion of the hearing, the committee decided to recommend a "wait and see" response, since the process for the development of the regulations involved several further opportunities for input by industry.

In developing the regulations, the Interagency Task Force issued two discussion papers, held two large public workshops, and held several small meetings with industry groups, hazardous waste management firms, and environmental groups. As the process continued, the issue became not whether restrictions on land disposal were needed or would be imposed, but what the restrictions would look like. Several changes were made in the initial proposals concerning the categories of hazardous wastes to be restricted, the dates for the restrictions, categorical exemptions for certain wastes, and the variance procedure. By the time the formal process for adopting proposed regulations had begun, most groups agreed with the need for restrictions on land disposal and with the general outlines of the process being proposed, although specific disagreements remained.

The final package of regulations represented several compromises. Land disposal of sludges containing toxic metals was not restricted as originally proposed. Restrictions on the broad class of hazardous wastes called extremely hazardous wastes was deferred until the class could be better defined by other regulations being developed. Concentration thresholds were set somewhat higher than originally proposed. And the proposed restrictions on volatile organic wastes were withdrawn so that the problem of smog-producing emissions could be dealt with by the Air Resources Board. The Interagency Task Force felt that these compromises were necessary to get the program into place and to establish the precedent for restrictions.

PROGNOSIS FOR THE SUCCESS OF THE PROGRAM

So far the program has had the desired impact in California. Even before Governor Brown issued his executive order, hazardous waste treatment firms began approaching state officials with proposals for new recycling and treatment facilities. The program also has the public support of the new governor, George Deukmejian. Although he frequently disagreed with Brown's environmental programs, Governor Deukmejian stated shortly after his election that, "I support development of state policies to reduce dependence on land disposal of hazardous wastes in favor of safer forms of disposal, such as recycling or treatment facilities."[12]

Based on the results of a recent hearing, the Department of Health Services concluded that there will be no problem with the availability of treatment capacity for cyanide wastes restricted since June 1. Two facilities will have permitted cyanide treatment capability: the IT Corporation treatment facility in the San Francisco Bay area and the Casmalia Resources facility near Santa Barbara. Another firm, Chemical Waste

Management, has scheduled a cyanide destruction unit for operation by the end of August 1983.

It is also anticipated that the second phase of restrictions will go smoothly. The IT facility is capable of treating toxic metal wastes and strong acids, and the BKK Corporation, the owner of a large land disposal facility in the Los Angeles area, has one small treatment facility operating now in San Diego and a major one in preliminary stages of construction near Los Angeles. PCB liquids are already being treated in the state by the PCBX process, a portable dechlorination process that detoxifies the wastes, leaving a reusable oil.

The final two phases of the restrictions are less certain at this time. Recycling facilities for halogenated solvents are already operating in the state, but a full ban on the land disposal of wastes containing halogenated organics will also require permission for and construction of thermal destruction facilities to burn those wastes that cannot be recycled. An operating cement kiln owned by General Portland, Inc., and located in Lebec (near Bakersfield), has been approved to burn waste solvents with less than 2 to 3 percent halogenated organics. Other facilities have been proposed: the IT Corporation has recently applied for a permit to test burn halogenated organics in a cement kiln in a remote location east of the San Francisco Bay area. Chemical Waste Management is also planning to construct an incinerator at their Kettleman Hills land disposal facility, and the BKK Corporation has expressed interest in incorporating incineration into their new treatment facility. The approval process for a major hazardous waste incineration facility is much more rigorous and lengthy than that for small treatment facilities; this may delay the final phases of the initial restriction schedule.

APPLICABILITY TO OTHER STATES AND THE FEDERAL GOVERNMENT

California's initiatives were preceded by actions in other states to encourage the construction of facilities to recycle, treat, and destroy hazardous wastes. New York, Louisiana, Texas, and Massachusetts each have identified the type of facilities they want to encourage in their states and have taken steps toward their creation. California borrowed heavily from work done by these states. Yet the measures California has taken to discourage land disposal and encourage alternatives are considerable contributions. These are all the more exemplary since, as a major generator of hazardous wastes, California can still provide inexpensive land disposal capacity for virtually all its hazardous wastes. This refusal to pursue the path of dumping has set a model appropriate for the many states now suffering shortages in landfill capacity.

Other states have begun to place restrictions on land disposal. New

York failed to pass legislation creating a program of restrictions but has required land disposal operators in the state to plan for a shift to alternative technologies before any more permits will be granted. Illinois passed legislation to require the use of alternative technologies for all hazardous wastes starting in 1987, when their use would be technically feasible and economically reasonable. The state of North Carolina, which has no permitted commercial landfills, is now considering legislation that would require nearly all hazardous wastes to be rendered non-hazardous before landfilling can be considered.

The problem with this piecemeal approach to land disposal restrictions is that hazardous waste generators can always transport their wastes to a state where restrictions do not exist. This will not be such a problem in California due to the expense involved with the long transportation distances and the fact that wastes will not be restricted unless alternative technologies are in place. In eastern states, however, the transportation distances to less restrictive states are not prohibitive. The only way to avoid this problem is for land disposal restrictions to be imposed on the federal level.

Interest has been aroused in such a program on the national level. U.S. Senator Gary Hart and Congressman James Florio both were supporters of the California program during its development. In a letter of support, Senator Hart stated:

> Because the problem of hazardous waste landfilling is not unique to California, I believe it is necessary to adopt this program nationwide. It is extremely important that we as a society begin to explore and adopt positive solutions to the problem of hazardous wastes and I believe the California program is a significant and needed step in this direction.[13]

Legislation has been proposed that will require the U.S. EPA to begin a program of restrictions on land disposal based on the California program. Shifting to the use of recycling, treatment, and destruction technologies to permanently manage those wastes that are least appropriate for land disposal is the only way to avoid postponing the problem to the future.

NOTES

1. Congress of the United States, Office of Technology Assessment. *Technologies and Management Strategies for Hazardous Waste Control.* Washington, D.C., March 1983.
2. California Administrative Code, title 22, division 4, chapter 30, section 66900, *et seq.*
3. California Health and Safety Code, section 25175.
4. California Health and Safety Code, section 25174.

5. California Health and Safety Code, sections 25300, *et seq.* In fiscal year 1981–1982, this fee averaged about $8 per ton.

6. California Health and Safety Code, sections 25160, *et seq.*

7. 42 United States Code, sections 6901, *et seq.* 1976.

8. California Administrative Code, title 23, chapter 3, subchapter 15, sections 2500, *et seq.*, 1972.

9. Stoddard, S. Kent; Gary A. Davis; Harry M. Freeman. Toxic Waste Assessment Group, Governor's Office of Appropriate Technology. Sacramento, Calif. October 1981.

10. Chau, Pao C.; Dan Coffey; and David F. Ollis. *Off-Site Hazardous Wastes in California: Generation and Disposal Patterns.* Davis, Department of Chemical Engineering, University of California, 1981.

11. Executive Order B–8881, October 13, 1981.

12. Harrison, Donald. "The Deukmejian Dilemma: Promises Meet Reality," *California Journal*, December 1982.

13. Letter, Senator Gary Hart to Ron Wetherall, California Department of Health Services, October 20, 1982.

7

Preventing Groundwater Contamination: The Role of State and Federal Programs

SHEILA BROWN

In the late 1970s groundwater was often called the "sleeping giant" of environmental concerns. In 1983 the giant is fully awake. Groundwater resources are seriously threatened by improper waste disposal, indiscriminate dumping, and overuse. No other environmental protection effort will more sorely test the ability of state governments to deal with a public health and environmental menace.

In coming to grips with the impact of degraded water supplies, states will have to make difficult decisions regarding scarce budget resources and personnel. Even more troubling will be the overturning of conventional notions of groundwater rights and an extension of the federal government's role in protecting this resource.

Some states, such as New Jersey, New York, and California, have already initiated relatively sophisticated groundwater protection systems. Other states have only begun to evaluate their groundwater resource, while still others show little interest in taking even that first step. Yet no state is unaffected by groundwater pollution. Even in those few states that do have a comprehensive system, significant questions are raised about the effectiveness of the programs and the level of protection provided. There is a need for the federal government to give coherence and direction to the states in their efforts to arrest groundwater contamination. The aim would be to assist those who are already attempting to deal with the problem and to prod those who are recalcitrant in facing the issue.

A NATIONAL PROBLEM

Groundwater sources do not honor state boundaries. Surface water in one state may feed a recharge area for an aquifer in another. States can no longer deal with groundwater in isolation: economic interdependence and hydrogeologic reality make that ineffectual. The extent of this country's dependence on groundwater, and the increasing incidence of groundwater pollution, suggests a problem of national dimensions.

Almost all experts on groundwater note with dismay the disturbing lack of data on contamination and its ecological and human health effects. It has only been within the past five years that the scientific and regulatory community has even perceived a need for such information. As the results of these surveys accumulate, the problem assumes a clearer focus, although its full dimensions remain elusive.

Data from EPA's ten regional offices reveal that virtually every state has abandoned some drinking water supplies because of groundwater contamination.[1] The following examples illustrate the pervasiveness of the problem:

- A draft policy now under consideration by EPA states that about 3 percent of public water systems using groundwater show levels of volatile organic chemicals in the range of possible health concern. Nearly 2,000 public water systems may now be adversely affected.[2]
- Testing the water supply of 39 cities, another report revealed that some solvents were present in well over one third of the cities' supplies. For example, trichloroethlyene (TCE), a widely used solvent and degreaser, was found in 38.5 percent of the municipalities' raw groundwater supplies.[3]
- Another EPA survey, called the "1000-City Survey," analyzed samples from 500 cities chosen at random and 500 chosen by the states: 28 percent of the state-chosen sites and 22 percent of the ones chosen at random were contaminated, primarily with volatile organic chemicals.[4]
- Rural areas are not exempt. A joint study by Cornell University and EPA found that at least 67 percent of rural household drinking supplies carry at least one substance in levels higher than considered acceptable under the federal Safe Drinking Water standards.[5]
- A New Jersey study of 670 wells completed in 1981 reported that 16.6 percent of the wells showed some volatile organic contamination greater than 10 parts per billion; 3.1 percent showed contamination greater than 100 parts per billion. Although the state said 95 percent of the wells were safe for drinking, 31 wells were contaminated and 11 closed.[6]

The effects that these chemicals have on human health and ecological systems is the cause of major concern and public alarm. Synthetic organic chemicals can cause both acute and chronic health effects. TCE, for example, the solvent found so often in water supplies, has produced cancer in laboratory mice.[7] For many substances there is simply no information available on health or environmental effects.

Awareness of the groundwater pollution problem has grown concurrently with the discovery of thousands of toxic waste dumps strewn across the country. Landfills and industrial waste pits, ponds, and lagoons are a continual source of groundwater contamination. An EPA report, leaked to the press by a U.S. Congressman, indicates a total of 180,973 such facilities used to impound waste liquids. Most were chosen as sites without regard to groundwater quality. The Agency estimates that 90 percent of these may threaten groundwater.[8] When the EPA published its list of the 418 waste sites considered the most dangerous, 347 posed direct threats to drinking water supplies.[9]

The rise in reported incidents of groundwater contamination comes at a time when this country is becoming far more dependent on groundwater. The nation's pattern and rate of use has changed markedly in the past three decades. The rate of groundwater withdrawals in the U.S. has increased on an average of 3.8 percent a year since 1950, almost twice the rate of increase of surface water usage.[10] Industry and agriculture are heavily dependent on groundwater,[11] and fully one half of the population relies on it for drinking water.[12] Ninety-five percent of rural Americans obtain their water from wells. EPA estimates that groundwater withdrawals for all uses are growing at a rate of 25 percent a decade.[13] Groundwater does replenish itself but often not at the rate it is withdrawn. For example, in the Southwest and Great Plains areas, 27 of 45 subregions are experiencing overdraft.[14]

In some instances, depletion can lead to contamination. Pollution will spread more quickly through an aquifer that is pumped frequently. Groundwater withdrawal or overpumping may cause subsidence, altering the natural hydrogeologic flow or reversing the flow and inducing saltwater intrusion.[15]

In addition to depletion and leaking landfills, other major sources of contamination include agriculture, petroleum and natural gas production, mining, and underground injection. As might be expected, regions with considerable industrial development and high population density report that industrial wastes and domestic sewage have the greatest impact on groundwater, while the western and southwestern regions are more commonly concerned with the effects of mining and petroleum operations.[16]

EPA has not yet completed a congressionally mandated study of waste drilling mud and brines related to petroleum production. Nevertheless, 17 states already have documented cases of groundwater contamination from this source. Thirty million barrels of brine are disposed of every day in the U.S. by underground injection, another potential source of contamination.[17]

Once contaminated, groundwater may be impossible to clean. It certainly will prove expensive. Groundwater moves very slowly and it may take many years for pollution entering at one location to appear in a water supply somewhere else. Once discovered, testing and monitoring procedures are very costly, requiring sophisticated instruments.

One of the more surprising and disturbing revelations of the past decade has been the extent to which groundwater is vulnerable to contamination. Filtration and absorption simply cannot fully cleanse toxic organic chemicals from a water supply and many substances, such as common chlorinated hydrocarbons like PCBs, cannot be broken down by plants and microorganisms.

How are we as a nation to deal with this problem? Do the states have the resources, the technical know-how, and the political will to protect us from the harmful effects of contamination? Do we need a federally mandated program to impose some minimum requirements on the states to guard against further degradation?

DIFFICULTIES IN STATE CONTROL

Groundwater has historically been a matter of state governance. Until recently, state groundwater programs revolved primarily around issues of water rights and allocation problems. The federal government's role had been limited to gathering information and providing data to the states on such items as depletion rates and aquifer locations, as well as providing financial assistance for water projects.[18]

The discovery of serious groundwater contamination has prompted most states to pass minimum regulatory programs to protect the resource.[19] Often, responsibility for administering these laws is scattered among several state agencies and no state has enough personnel or resources to properly carry out an effective program.[20] Despite the documented danger presented by the 180,000 waste impoundments around the country, only six states have developed regulations that specifically address groundwater contamination from surface impoundments.[21] Fifteen to 20 states have laws in place or are developing laws to classify aquifers by use,[22] and some have groundwater quality standards for specific substances. But implementation of these state programs has been uneven, for states often fail to enforce these laws.

There is no evidence that even those states with sophisticated groundwater control systems have prevented further degradation of their water supplies. New York, for example, has a system that classifies groundwater and sets groundwater quality standards. Nonetheless, significant pollution consisting of synthetic organic chemicals is present in the water supplies of Nassau and Suffolk counties and many wells have been closed.[23]

New Jersey is considered to have one of the more advanced groundwater regulatory programs, including a classification system and water quality standards for each class.[24] Yet the state suffers from incident after incident of well closings from contamination. New Jersey has the highest number of waste sites on the national Superfund top priority list of any state, virtually all of which are there because they threaten the drinking water supplies of some community.

Although many of these sources of contamination were present prior to the passage of these state laws, the states' capability to contain these pollution sources and prevent further deterioration of groundwater quality will become the central issue in the next few years. Many observers argue that states can not deal successfully with this problem without substantial assistance from the federal government.

This is because groundwater contamination presents a state with a whole new array of technical, political, and economic difficulties. Groundwater monitoring and testing is scientifically complex and expensive. It requires trained, experienced professionals who are often in short supply. Constricting state budgets and an anti-bureaucracy political climate further impede a state's ability to maintain adequate protection programs. Locally based industries which discharge wastes often have a significant political influence on state legislatures sensitive to threats of plant closings and job losses. The result is that few states have attempted to develop groundwater quality data in a comprehensive way.

In 1980, an EPA official testifying before Congress said that protection of groundwater supplies is the responsibility of states, but few have the necessary authority to do so and most do not recognize there is a problem.[25] As time goes on, increasing numbers of states recognize the problem and many are passing laws creating additional authority to deal with it. But it appears increasingly unlikely that states alone will be able to generate the determination and resources to successfully prevent widespread degradation of water supplies. What is likely to emerge is some sort of a hybrid system that contains both national and state elements. It must be flexible enough to allow for state differences, yet uniform enough to avoid national confusion.

A LEGACY OF FEDERAL INACTION

Those who hope that the federal government will be the white knight, rescuing water resources for the states, can take little comfort from the federal government's record to date. As environmental legislation poured forth from Washington in the 1970s, groundwater considerations were dealt with only in a piecemeal and uncoordinated fashion. In fact, groundwater is the only remaining major environmental medium that lacks a comprehensive federal regulatory scheme. Groundwater escaped notice for numerous reasons. Until recently, groundwater was thought to be largely pristine, capable of cleansing itself, and of nearly endless supply.

States were generally given deference on groundwater issues, and the federal government remained relatively uninvolved.

To the extent that it has been involved in groundwater problems, Washington's response has been largely "piecemeal, fragmented and incomplete."[26] There are, in fact, several statutes through which the federal government could have become more actively involved in at least some aspects of groundwater regulation.

The Federal Water Pollution Control Act gave EPA power "to prepare or develop comprehensive programs for preventing, reducing or eliminating the pollution of navigable waters and groundwaters."[27] But confusion over the regulatory provisions of the Act, which speak only of "navigable waters," had led EPA to focus its efforts exclusively on surface water matters. To date, the Agency has not sought to clarify or exercise any authority over groundwater.

The Safe Drinking Water Act,[28] passed in 1974, is the only federal statute designed to deal directly with water quality. The Act's narrow mandate calls for federal standards for publicly supplied drinking water and for protection of sole source aquifers. Congressional committees have held hearing after hearing in exasperation over EPA's seeming inability to implement this Act. Very few standards have been set and only a handful of aquifers designated as sole sources. EPA has complained about the scientific and technical difficulties involved in setting the standards. According to a recent study published by Princeton University the demise of the Act is obvious:

> The water supply industry has temporarily blunted the effectiveness of the Safe Drinking Water Act by fighting implementation of many regulations under the Act, particularly those designed to reduce synthetic organic compound contamination. Much of the forward momentum in implementing the Act that occurred shortly after it was passed in 1974 has been lost. Only a limited number of synthetic organic chemicals are regulated; the frequency of monitoring requirements has been constrained; government aid has been lacking for small water purveyors that have insufficient resources to meet standards and institutions have been slow to hire more and better trained personnel needed to comply with the more sophisticated requirements of controlling chemical contamination.[29]

The Resource Conservation and Recovery Act (RCRA), which was designed to cover the management of solid and hazardous wastes, can address the problem of contamination from land disposal in the future by requiring strict performance and technical standards. But the Act is only now being implemented and its regulations continue to undergo revisions. Standards proposed so far clearly embody compromises that make it uncertain whether RCRA can adequately protect groundwater. Also, since RCRA covers only certain listed or described wastes, contamination from other sources will not be covered in the regulatory scheme.

The Comprehensive Environmental Response, Compensation, and Liability Act, or "Superfund," was enacted in 1980 to clean waste dumps and toxic spills. Although water contamination is the most serious problem resulting from these waste sites, EPA has refused in its general plans to set definitive standards for restoration of water supplies. The Agency has failed even to require remediation to levels equal to those few existing water quality standards under the Safe Drinking Water Act.

A few other federal statutes treat groundwater tangentially, such as the Federal Insecticide, Rodenticide, and Fungicide Act; the Toxic Substances Control Act; the Surface Mining Control and Reclamation Act; and the Uranium Mill Tailings Act. All of these Acts are directed at specific areas of regulation, as their titles reveal, and each only incidentally addresses effects on groundwater.

While many of these statutes overlap, all of them taken together do not provide a comprehensive protection program. As one congressional committee concluded, "To date the Federal effort to control ground water pollution has been haphazard and ineffective."[30] The problem is caused by a number of factors. The piecemeal approach has created a patchwork of laws, difficult for EPA and other agencies to implement. The great void indicates a growing need for a national program, comparable to that for surface water, to classify groundwater by designated uses, to establish groundwater quality criteria, and to implement an enforcible groundwater protection program.[31]

THE RISE AND FALL OF THE FIRST FEDERAL STRATEGY

In 1979, the federal government was indeed taking note of groundwater contamination. That summer, a task force appointed by President Carter to study groundwater completed its report and concluded that the country was at a crossroads with respect to the welfare and future of its groundwater resources. In addition to recommending the creation of a special commission to analyze groundwater policies, the report stated flatly that existing water-planning measures are inadequate and that guidelines for coordination of state and federal government activities are needed.

A special commission was not created, but EPA began work immediately on a federal groundwater policy. Soon after, the administration's budget reflected its awakening concern by requesting new funds for a groundwater protection program.[32]

In November 1980, shortly before President Carter left office, EPA published its *Proposed Groundwater Protection Strategy.* As an initial policy statement, the document is quite alarming. It proclaims the current groundwater problem is becoming unmanageable and warns that states and localities might soon be overwhelmed by the growing instances of contamination.[33] Calling groundwater contamination a "critical national imperative," it asserts that public institutions are not now well equipped to respond effectively or efficiently to the problem.[34]

Despite the intensity of its warning, the proposed strategy is modest. It envisioned a coordinated partnership between the federal government and states whereby groundwater protection strategies would be developed by all states with guidance from EPA. Groundwater would be classified according to use, and the use would determine the level of degradation which would be permitted. Meanwhile, EPA would continue to implement its existing statutory authorities and determine whether additional legislative action was needed. The strategy recommended that states continue to have the key responsibility for groundwater management and protection with the federal government providing guidance, setting classification standards, and doing research and development and enforcement. The federal government might intervene directly only when a groundwater threat was of such high priority as to warrant strong national action or when another federal statute was called into play.

Comments on the strategy were almost universally favorable, all parties agreed on the need for a federal policy on groundwater. There was some disagreement over specific provisions. For example, the Chemical Manufacturers Association (CMA),[35] which represents the industry that produces the largest amount of hazardous waste of any sector, felt that the proposed strategy taken as a whole amounted to a veiled attempt to impose a "non-degradation" standard on the nation. A strict non-degradation standard policy would not allow activities that would further degrade a water supply below its present quality. The CMA reached this conclusion primarily because of EPA's stated intention to protect all water supplies currently at drinking water quality, although the strategy clearly stated that this would be the policy only until a classification system was in place.

Another industry representative objected to EPA's use of the word "enhance" in stating the strategy's objectives to protect and enhance groundwater supplies. The representative feared, it seems, that this might engender some remedial costs in upgrading despoiled existing supplies.[36] Industry commentators in general were concerned about two main issues: that EPA not promulgate a non-degradation policy and that the federal government not interfere with or preempt the states' traditional role in groundwater management.

Environmental groups, on the other hand, voiced skepticism that states would "willingly assume their assigned role in the program" and suggested that there might well be strong resistance, especially in states where groundwater control has been minimal.[37] The Environmental Defense Fund supported the strategy, stating that the continued loss of groundwater resources is a national problem requiring a national effort.

The policy must allow for state differences but there must be a strong federal component focusing on technical guidance, research and development, enforcement, regulatory controls over toxic materials, standard setting, and provision of interstate consistency where needed.

EPA held five public hearings around the country on the strategy in early 1981 which generated quite a bit of interest. Over 1,000 persons attended the hearings, 172 testified, and 309 submitted written comments, generally of a favorable nature.

The strategy was temporarily forgotten amidst the confusion accompanying the change of administrations. But a hint of things to come was present in President Reagan's first budget to Congress which eliminated $7 million previously earmarked for state groundwater management grants and for the development of state groundwater plans.

In September of 1981 a Connecticut Congressman began a series of correspondence with new EPA Administrator Anne Gorsuch concerning the delay in implementing the strategy and requesting the new administration's views on groundwater policy. The administrator initially responded that she had not yet had time to focus on the best approach to groundwater issues but that it was a topic of primary concern. She also stated that no EPA funds or staff would be specifically earmarked in the budget to work on the strategy but that staff from other areas at EPA would work on it.

The administrator subsequently ordered her staff to revise the original EPA strategy, and a year later, in October of 1982, EPA did release a new draft groundwater policy.[38]

The new policy contains a dramatic change of emphasis. It stresses emphatically the voluntary nature of states' participation and reaffirms their primary role in groundwater protection. Implementation of protection strategies will depend entirely on each individual state's desire or ability to coordinate its own efforts. EPA's role will be merely to coordinate existing laws and federal activities. It flatly rejects any notions of non-degradation of groundwater and embraces the idea that the degree of protection to be sought depends upon the use to be made of the particular supply. The new strategy would rely entirely on existing statutory authority and would seek no new legislation, or even any new program structure at EPA. It is surprisingly pessimistic about ever obtaining the kind of scientific data needed for a complete understanding of groundwater pollution, stating, "There is no feasible way to determine comprehensively, through monitoring, the full extent of groundwater contamination on a national basis."[39]

The new draft strategy is criticized by congressional staff for emphasizing state roles at the same time the administration is seeking to cut grant programs from the budget that the states rely upon for groundwater protection. They also object to the strategy's lack of a policy for preventing degradation of all groundwater, lack of a national groundwater classification system, and lack of a national system for monitoring groundwater.[40] One EPA official has commented that the policy as now written is ineffectual because "it doesn't commit a lot of resources, doesn't make anyone do much of anything and doesn't give states anything."[41]

Despite the moderate, even innocuous, nature of EPA's present draft, it has run into strong opposition within the Reagan administration and has not yet been adopted as official policy. Former Secretary of the Interior James Watt, who chaired the Cabinet Council on Natural Resources and Environment, believes that the groundwater strategy signals the federal takeover of states' rights and held up action on the policy while he tried to gather support for his position. He believes that the federal government's role should be limited to providing information to the states on what other states are doing on groundwater.

On February 25, 1983, Secretary Watt sent a letter to the western governors saying there was a "movement brewing in Washington to establish federal control over ground water" and asking them to join him and voice their opposition to the proposed policy. Apparently the western governors do not share Mr. Watt's concern about a federal takeover of their rights since three of the interstate groups representing the governors, including the Western State Water Council, the National Governors Association, and the Interstate Conference on Water Problems all have rejected his plea. All three groups generally support EPA's national strategy. While obviously wanting to retain state primacy over groundwater management, they see a need for a federal presence in many areas.[42] The draft policy is still awaiting action at the Cabinet Council.

THE NEXT STEPS

Whatever the political persuasions of those who hold state or federal office, they will have to respond to the groundwater question in the next few years. Public awareness of the issue is growing rapidly as new evidence of contamination is found and new well closings are announced. A Lou Harris poll released in December of 1982 reported that 88 percent of the public believes that groundwater contamination is a "serious national problem." With those kinds of numbers, political institutions can no longer afford to ignore the issue.

The draft groundwater strategy proposed by EPA in 1980 said, "Managing instances where contamination is discovered to avoid future public health problems is a major task for the future which is beyond the past experience of most units of state and local governments." Some states, like New Jersey, Connecticut, and New York, have begun their own classification systems, but many states are waiting to see what action EPA takes on its proposed groundwater policy. Further, it is not at all clear that all states yet perceive groundwater contamination as a major problem. In the absence of additional federal action, states will be forced to protect their resources and choose among a myriad of policy options to control groundwater usage and pollution.

State activity will run into a full complement of impediments. States are desperate for funds and for technical expertise. Competition with other state programs for dwindling budget resources will be fierce. Without continued federal funding only highly visible environmental programs will be carried out at high levels of commitment. Vital but unglamorous items like research, planning, and data gathering may fall victim to lack of resources, and states may only be able to address the most urgent problems in a stop-gap manner. Any attempt by a state to aggressively control future groundwater contamination can expect to run head on into powerful economic interests who fear additional costs and resent additional regulation.

The irony of the states' dilemma is that whether they proceed vigorously or sit back and do nothing, either course is likely to lead to increased federal involvement. Certainly, if the states fail to take any action, whether from lack of resources or lack of will, the public pressure will grow on Washington to address the problem. On the other hand, the proposition of having 50 different sets of state water quality rules may demand a federal system coordinating the state systems.

Senator Robert Stafford (R.-Vt.), Chairman of the U.S. Senate Committee on Public Works and the Environment, has said in relation to environmental protection that "relying too much on state and local efforts could recreate the very problems we sought to avoid three decades ago."[43] The problems he spoke of were brought about by a proliferation of local laws which prompted industry, especially the chemical and auto industries, to push for consistent, uniform regulations. It was this initial industry pressure, linked with state and local governments' concern about economic equity, that lead to increased federal presence in environment protection over the past several years.

If these pressures again assert themselves, Congress will have to decide the shape and content of a federal groundwater program. Congress has already demonstrated its considerable interest in the subject. In 1982, the House passed a bill creating a National Groundwater Commission to study problems of toxic contamination and overdraft.[44] Though never adopted by the Senate for other reasons, there were overwhelming expressions of support by members of both parties for the commission idea and for the need to address groundwater problems.

Clearly, any federal program will have to take into account the extraordinarily sensitive nature of groundwater issues because of their historic association with states' rights. There is likely to be a strong resistance to additional controls by either the state or federal government, especially in some western states where water is a constitutionally protected property right. But states now support at least a limited federal presence in terms of financial aid, technical assistance, and research and development. Thus, many recognize the need for a fuller federal program.

Even assuming that the gaps in groundwater protection can be filled by a fuller implementation or more liberal interpretation of existing statutes, the fact remains that until groundwater is made a legislative priority by the Congress, EPA's resources and energy will continue to be directed toward the major change of existing statutes. For this reason it may take a specific, focused statement of congressional intent, through a groundwater bill, to redirect EPA's planning and budget.

If Congress decides to take up groundwater legislation it will not have an easy time. In the House there are at least four full committees that are likely to have an interest and will attempt to assert jurisdiction over any bill that affects groundwater. Powerful interests will seek weak provisions and special exemptions. But Congress has faced all of these before and has managed, in other environmental areas, to establish reasonably protective regulatory programs.

Components of a federal bill might include guidelines on the classification of water supply uses, standards for protection of those uses, research and development, and technical and financial assistance to states. The information gap must be addressed with requirements for regular monitoring and data gathering. Some have suggested that legislation is needed to create incentives for long-term source reduction, which could involve new taxes as well as tax credits, restructured liability, use of market mechanisms to encourage recycling, waste reduction, and process modifications. Land use planning is recommended by many as a prime mechanism for protecting groundwater but mainly as a tool to be used exclusively by state and local authorities. Yet, the federal government is already imposing some land use requirements through Resource Conservation and Recovery Act regulations regarding the siting of new hazardous waste facilities.

Certainly there are some areas of a comprehensive groundwater program that are best left to the states. The legislative process has the capacity to assess and accommodate the delicate balance between state and federal prerogatives as legislation is hammered out. At the same time, it must be recognized that at least some elements of the groundwater contamination problem are beyond the states' ability to cope with or afford. Simplistic platitudes about states' rights and a domineering federal government will not bring this country any closer to a safe and healthy water supply. What is needed, instead, are comprehensive federal standards.

A recent report by Princeton University compared American and European approaches to groundwater pollution.[45] Realizing that it is probably impossible to obtain sufficient information to satisfy all questions concerning toxic chemicals in drinking water, the Europeans have taken a number of preventative measures. For example, granular activated carbon filters are in common use by water suppliers to reduce synthetic organic chemicals. Unlike Americans, Europeans have been willing to pay for such advanced control technology in exchange for the extra measure of protection it affords. The report is pessimistic about our ability to control contamination at its source and goes on to recommend land use controls, better-trained personnel, sophisticated and stronger monitoring, federal incentives for stronger state programs, and implementation of EPA's groundwater strategy with non-degradation as its goal.

Coping with the groundwater problem will challenge the capabilities of all levels of government. Because of the nature and history of this problem, success in dealing with it will come only if due respect is given to the differing abilities of each level of government. Neither trampling on states' rights nor categorically rejecting federal involvement will lead to success. The problem must be faced squarely and cooperatively by all levels of government. EPA's 1980 draft strategy sums up the current predicament. "The little we have learned in recent years has taught us three sobering lessons: the problem is real and dangerous; our present state of knowledge to deal with these problems is limited; and present programs, authorities and resources are inadequate."[46]

NOTES

1. Testimony of Robert Harris, member of the Council on Environmental Quality before the House Government Operations Committee, Subcommittee on Energy, Environment and Natural Resources, July 24, 1980.
2. 13 Ent's Rep. (BNA), no. 26 (October 29, 1982), p. 907.
3. U.S. EPA, Office of Drinking Water, *The Occurrence of Volatile Organics in Drinking Water* (U.S. Government Printing Office, Washington, D.C., March 6, 1980).
4. Speech by Joseph Cotruso, director of Criteria and Standards Division, Office of Drinking Water, U.S. EPA, to American Water Works Association (Hartford, Conn., June 3, 1982).
5. U.S. EPA and Cornell University, *National Statistical Assessment of Rural Water Conditions* (U.S. Government Printing Office, Washington, D.C., 1982).
6. Robert K. Tucker, *Groundwater Quality in New Jersey* (Trenton, Office of Cancer and Toxic Substances, N.J. Department of Environmental Protection, 1982).
7. Council on Environmental Quality, *Environmental Quality 11th Annual Report* (U.S. Government Printing Office, Washington, D.C., 1980), p. 92.
8. U.S. EPA, *Surface Impoundment Assessment: National Report*, released by Congressman Toby Moffett (D.-Conn.), (U.S. Government Printing Office, Washington, D.C., December 29, 1982).

9. Office of Technology Assessment, *Technologies and Management Strategies for Hazardous Waste Control* (U.S. Government Printing Office, Washington, D.C., March 1983), p. 5.

10. Conservation Foundation, *State of the Environment, 1982* (The Conservation Foundation Press, Washington, D.C. 1982), p. 108.

11. U.S. Water Resources Council, *The Nation's Water Resources 1975–2000*, vol. 1 (U.S. Government Printing Office, Washington, D.C., 1975), p. 13.

12. CEQ, *Environmental Quality*, p. 84.

13. H.R. Rep. No. 96–1440, 96th Cong., 2d Sess., p. 2.

14. U.S. Water Resources Council, *Groundwater Overdraft by Region* (U.S. Government Printing Office, Washington, D.C., 1975).

15. Tripp and Jaffe, "Preventing Groundwater Pollution: Towards a Coordinated Strategy to Protect Critical Recharge Zones," *Harvard Environmental Law Review*, vol. 3 (1979), pp. 1–47.

16. CEQ, *Environmental Quality*, p. 87.

17. Speech by John S. Fryberger, consultant, Engineering Enterprises, Inc., Norman, Okla., to American Water Resources Association (San Francisco, October 12, 1982).

18. Conservation Foundation, *State of the Environment*, pp. 113–14.

19. Ibid.

20. U.S. EPA, Office of Drinking Water, *Proposed Groundwater Strategy*, (U.S. Government Printing Office, Washington, D.C., November 1980), p. 13.

21. EPA, *Surface Impoundment Assessment*.

22. U.S. Water Resources Council, *State of the States: Water Resources Planning and Management, Groundwater Supplement* (U.S. Government Printing Office, Washington, D.C., 1981), pp. 1112–13.

23. Council on Environmental Quality, *Contamination of Ground Water by Toxic Organic Chemicals* (U.S. Government Printing Office, Washington, D.C., January 1981).

24. New Jersey Water Pollution Act, 58 N.J. Stat. Ann. Sec. 10A–1–10A–14 (1977).

25. Testimony of Eckhardt C. Beck, assistant administrator for water and waste management, before the House Government Operations Committee, Subcommittee on Energy, Environment and Natural Resources (July 25, 1980).

26. Speech by Russell Peterson, president of the National Audubon Society, former chairman of the Council on Environmental Quality, to Water Conference of the Northeast-Midwest Coalition, (Washington, D.C., February 1982).

27. 33 U.S.C. 1252(A).

28. 42 U.S.C. 300f–300j–9.

29. Princeton University, Center for Energy and Environmental Studies, *Nor Any Drop to Drink: Public Policies Toward Chemical Contamination of Drinking Water* (Princeton, N.J., 1982), as reported in 13 Env't. Rep. (BNA), no. 19 (September 10, 1982), p. 643.

30. H.R. Rep. No. 96–1440, p. 4.

31. Tripp and Jaffe, *Preventing Groundwater Pollution*, p. 25.

32. Speech by Morgan Kinghorn, acting chief, Office of Management and Budget-Environment Branch, to Association of State and Interstate Water Pollution Control Administrators (Washington, D.C., January 1980).

33. EPA, *Groundwater Strategy*, p. 3.
34. EPA, *Groundwater Strategy*, p. IV–1.
35. Chemical Manufacturers Association, *Comments on EPA's Proposed Groundwater Protection Strategy*, public docket (Washington, D.C., February 1981).
36. Letter from Dr. F. W. Chapman, Jr., manager, environment and energy conservation, Atlantic Richfield Corporation to Mariah Mlay, EPA associate deputy assistant administrator for drinking water (February 13, 1981).
37. Environmental Defense Fund, *Comments on EPA's Proposed Groundwater Protection Strategy*, public docket (January 1981), pp. 36, 38–39.
38. 13 Env't. Rep. (BNA), no. 26 (October 29, 1982), p. 907.
39. Ibid.
40. Ibid., no. 37 (January 14, 1983), p. 1577.
41. Ibid., no. 41 (February 11, 1983), p. 1800.
42. Ibid., no. 46 (March 18, 1983), p. 2084.
43. "New Federalism, Environmental Protection: Are They Compatible?" *Chemical & Engineering News* (May 10, 1982), p. 15.
44. H.R. 6307, 97th Cong. 2d Sess. (1982).
45. Princeton University, *Nor Any Drop to Drink.*
46. EPA, *Groundwater Strategy*, p. 2.

8

Europe's Detoxification Arsenals: Lessons in Waste Recovery and Exchange

BRUCE PIASECKI

While the debate rages in this country about the best ways to dump the millions of tons of toxic wastes generated each year, industrialized countries in Europe are quietly managing their wastes in facilities that resemble chemical plants rather than the dumps that have often haunted Americans.

As much as ten years ago, the land-conscious Europeans decided that dumping was neither safe nor economical for a wide range of conventional industrial wastes. On February 11, 1976, the Netherlands passed a Chemical Waste Act[1] which explicitly prohibited the dumping of a wide range of toxic wastes. This ban has spawned a thriving waste detoxification business: chemical, physical, and biological treatment techniques have been employed ever since, as well as innovative waste reduction, recovery, and recycling strategies.

While the majority of U.S. waste exchanges continue to report exchanging below 20 percent of what they list for recycling,[2] the Dutch successfully exchanged 30 percent of their listings as soon as the Act was passed. Throughout the mid-1970s, the small but crowded Netherlands listed 100 to 155 wastes annually.[3] In contrast, the first U.S. exchange, the Midwest Industrial Waste Exchange, exchanged less than 10 percent

This chapter originally appeared in the August 1983 *Northeast Industrial Waste Exchange Listings Catalog* (Issue No. 9) and is reprinted with permission of the Northeast Industrial Waste Exchange, Syracuse, New York.

The assistance of Gary A. Davis in the preparation of this piece is gratefully acknowledged. Davis, a hazardous waste specialist and lawyer, has researched European hazardous waste policies under the auspices of the International Institute for Applied Systems Analysis in Laxenburg, Austria.

of their 60 listings in 1979. Today, failures to improve the Resource Conservation and Recovery Act continue to hurt American efforts at exchange.[4]

Without the luxury of imagining land disposal to be limitless, the Netherlands reports that the pressure for resource recovery is now very great. An eminent Dutch researcher, Frits van Veen, noted before a recent NATO conference that there will be no more Dutch research on land disposal—instead, efforts will concentrate on the recovery of wastes.[5] While the bulk of U.S. funds continue to be sunk in attempts to secure landfills, the Dutch government actively supports this trend toward treatment and recycling, especially by supporting research on recovering heavy metals from metal sludges and from municipal waste incinerator fly ash.

Moreover, the same 1976 Act has a clause to prohibit the manufacture of Dutch goods that generate wastes "impossible or very difficult to dispose of" or that are "not properly stored, treated, processed or destroyed."[6] In contrast to the grueling process by which American policymakers learn to restrict the production of substances like DDT, PCBs, or dioxins, the language of this law gives Dutch officials an ability to preempt problems categorically. Having itemized the hazards of dumping in their waterlogged country, the spirit of this law inspires product substitution and waste reduction wherever possible.

Before dismissing this example as atypical, consider one thing: detoxification of hazardous waste and tenacious strategies of waste reduction and exchange are standard operating procedures in at least a half dozen European countries. (See notes 5 and 20 for citations.) The specifics, techniques, and strategies vary, but they all share three things: a willingness to stabilize treatment and recovery markets by regulatory reform, a knack at cooperative financing and joint management, and, most important, a profound distrust of land disposal.[7] A brief survey of Danish, German, Norwegian, and French facilities follows, in an effort to highlight common patterns of development and to illustrate the recurrent importance of restricting land disposal.

DENMARK

Danish national statutes, passed in 1973 and 1974, assign a dual responsibility for securing safe waste management. The community generating a waste must properly treat or dispose of the waste within the community's system. Should this prove impossible, the host community and the particular generator are required to send the difficult wastes to a central treatment facility—the Kommunekemi facility in Nyborg, the geographical center of the country.[8] Located on the island of Funen, this integrated facility recovers heat from its toxic waste incinerators and supplies 35 percent of the heat for Nyborg's 18,000 residents. The facility also includes neutralization, precipitation, and chemical destruction.[9] And if the Kommunekemi plant is not equipped to destroy or recycle a particular waste, the Danes don't dump it—they put it in storage until they find a way to treat it.

Kommunekemi was established by the creation of a nonprofit public corporation, Kommunekemi A/S, which was funded by low-interest government loans. (In contrast, since the U.S. IRS code prevents the use of municipal bonds to finance the construction of hazardous waste facilities, America continues to suffer from an acute shortage of treatment centers.) Owned by a consortium of municipalities, the Kommunekemi plant officials also operate 20 major collection substations so that no waste generator is more than 20 miles from a station. The wastes are then shipped by rail to Nyborg and housed in special cars also owned by Kommunekemi.[10] This shipment of waste is closely traced by a manifest tag system, not unlike those now required under the Resource Conservation and Recovery Act but more effectively enforced.

The director general of the Danish National Agency of Environmental Protection sits on the board of directors. The operation is audited twice each year by an independent lab, with complete tests for gas emissions, ash, and slag. Other aspects are tested weekly, and the precipitation runoff from the entire plant is collected and tested prior to release. According to John E. McClure, who visited the Danish facility for the Dames and Moore consulting firm, the entire operation of the plant is conducted in an open, full-disclosure manner. This approach, coupled with the benefit of subsidized energy to Nyborg, has created an atmosphere of acceptance by the community.

WEST GERMANY

The Netherlands and Denmark are not world-class industrial powers like West Germany. But those who doubt that integrated treatment and recycling can be made effective and economical need only visit any of that country's 15 waste treatment facilities, which are operated as part of a coordinated national program. Regulatory reform and innovation seem to work hand in glove. While Americans continue to dump close to 80 percent of their toxic wastes into the soil, almost 85 percent of all West Germany's wastes are sent to these 15 large treatment plants for destruction or reuse. The great degree of risk evident in this disparity is further heightened by the fact that Americans generate three times more toxic waste per citizen than do the West Germans.[11]

According to Dr. Berndt Wolbek from West Germany's Federal Ministry of the Interior, German legislation explains the country's shift toward treatment and recovery.[12] German law is written so that waste treatment disposal is foremost a responsibility of the government, although the duty can be transferred to private industry in select cases.[13] As a result, the German government attacks hazardous waste problems quite aggressively. Surveying the German example for the German Marshall Fund, Sandra E. Jerabek reports that by 1980 five out of ten German states had established "joint venture companies,"[14] cooperative arrangements between government and industry designed to arrest toxic contamination at its source.

The first of these companies, the GSB in Bavaria, exemplifies the ways creative financing can encourage the construction of costly high-tech facilities. With a stock of DM15 million, partners in this venture include the Bavarian state government, owning 40 percent; diverse city co-ops, owning 30 percent; and 50 district industrial corporations, owning the remaining stock. In response to these groups' needs, GSB operates whatever facilities deemed necessary in Bavaria, including a resource recovery facility, on a nonprofit basis. About the size of Ohio, Bavaria has found that three regional facilities can handle the bulk of its wastes, totalling 350,000 tons per year.[15] In contrast, New York State alone presently generates 400,000 tons of toxic waste beyond its present treatment capacity. Siting problems abound; what we can't treat, we dump.

The strong role the Bavarian government plays in moving beyond dumping is now clear. According to Gunter Eichele, an official working for the Bavarian State Ministry for Land Development and Environmental Questions, the Bavarian government funds 50 percent of most treatment and recycling efforts, but contributes about 20 percent for methods considered "less desirable," such as landfilling.[16] Although local governments can still decide on which technologies they will employ, the Bavarian investments significantly shape selections against dumping. The Ebenhansen/Gallenback complex near Munich, for example, reports that about 50 percent of its collected wastes are safely incinerated in rotary kilns.

The West German government also encourages extensive recycling efforts, since the state governments require waste generators to first show whether recycling methods have been explored before other methods can be permitted. According to Sandra Jerabek, the articulation of this priority has shifted many wastes once destined for incineration at sea into recycling pathways. Coordinated by the Federal Chamber of Industry and Commerce, the West German waste exchange boasts between 2,500 and 5,000 listings per year, with a 36 percent success rate of exchange.[17]

DECENTRALIZED CONTROL: THE FRENCH AND NORWEGIAN EXAMPLES

Unlike the large integrated facilities of Denmark and West Germany, the French prefer small decentralized facilities scattered across the country. In order to finance this approach, the National Ministry led a three-year search for appropriate sites. After sites were located, proposals for construction and operation were invited from private industry. In this way, E. Perol of the French Ministry of the Environment notes, the French government paid for the laborious searching and testing process—a cost that U.S. industries usually bear.

While American firms suffer from significant public relations and legal costs before siting a facility, the French approach inspired a series of group efforts—at times, even inciting unexpected cooperation from American firms in France. On the west coast of France near Sandouville, for example, 20 companies with oil and petrochemical plants in the area formed a consortium named the "Society of Regional Study for the Protection of the Environment" (SERPE) in 1973. This consortium, which included American-connected Ashland Oil, Esso, Firestone, Goodyear, Mobil, and Shell plants, coordinated the design and construction of a facility in Sandouville that recycles, treats, and destroys the majority of the wastes produced in the region. Day-to-day operations are carried out by an operator under contract, Sidibex, a subsidiary of Campagnie Generale Des Eaux.[18]

The Norwegians also take a decentralized approach, retrofitting existing industrial facilities for toxic waste destruction. Located along a wooded fiord near Oslo, the Norcem Cement Plant substitutes select toxic wastes for up to 30 percent of the fuel used to fire their cement kilns. The Norwegian government funded two-thirds of the retrofitting costs, requiring strict combustion standards. Other energy-intensive industries are expected to follow suit.[19]

These European examples haven't gone completely unnoticed in the United States. (See chapter 6 on the California example; also see the European Contact List which follows.) The lesson in all this is obvious but usually overlooked. The last thing Americans should do with toxic wastes is dump them untreated into our soils. Instead, we should follow the European example and do something safer and far more sensible: recycle what we can. What we can't, we should turn into harmless chemicals and make some money selling what's left over. Each step this country takes to restrict land disposal is a step toward waste recycling, treatment, and reduction.

EUROPEAN CONTACT LIST/SELECT FACILITIES

This contact list has been prepared with the help of Gary A. Davis for those interested in studying European facilities directly. It also contains brief descriptions of select facilities.

Austria

Austria has one of the most advanced treatment facilities in Europe, located just outside of Vienna. The EBS facility is now run by the city government. It has two large rotary kiln incinerators that include heat recovery for the district's heating system and generation of electricity. The incinerators also include sophisticated waste handling equipment so that liquid wastes can be automatically extracted from drums for pumping into the incinerator, and whole drums of solids can be fed in automatically on a conveyor belt. Unfortunately, the facility is not operating at full capacity because Austria lacks laws and regulations to force industries to use the facility for waste disposal.

Dr. Gerhard Vogel
Institut Fur Technologie Und Warenwirtschaftslehre
der Wirtschaftsuniversitat Wien
Augasse 2–6
A–1090 Wien
Austria
(0222) 34 05 25 806

Denmark

The Danes began planning their hazardous waste solution over ten years ago. At the heart of the Danish system is the Kommunekemi facility in Nyborg, the geographical center of the country. The facility is owned by a consortium of municipalities, who also operate a system of collection stations around the country, so that no hazardous waste generator is more than 20 miles from a station. Hazardous wastes are incinerated in large rotary kiln incinerators with heat recovery units supplying 35 percent of the heat for Nyborg's 18,000 residents. The facility also includes neutralization, precipitation, and chemical destruction.

Peter Lovegren
Managing Director
Chemcontrol A/S
Dagmarhus
DK–1533 Copenhagen V
Denmark
01–14 14 90

Finland

Thomas Aarnio
Ministerial Secretary
Ministry of the Interior
Environmental Protection Department
B.B. 257 SF–00171
Helsinki 17
Finland
0–160–4515

France

In order to finance their decentralized approach, the National Ministry led a three-year search for appropriate treatment sites. In this way, the French government paid for the laborious searching and testing process—a cost that U.S. industries usually bear. This approach was inspired by a series of group efforts. On the west coast near Sandouville, 20 companies formed a consortium in 1973. Members include Ashland Oil, Esso, Firestone, Goodyear, Mobil, and Shell plants.

Momed. Aloisi de Larderel
Chef du Serves des Dechets
Ministere de L'Environment
14, Bld Du General LeClerc
92524 Neuilli Cedex
France
1–758–1212

Netherlands

J. H. Erasma
Ministry of Health and Environment
10, Regersstraat
Leidschendan
Netherlands
(070) 209260

Norway

The Norwegians, who generate a much smaller quantity of hazardous wastes, have taken a different approach to creating the facilities to treat and detoxify them. They are creating a decentralized system using existing industrial facilities that are capable of treating wastes. For instance, the government has sponsored the use of a cement kiln at the Norcem cement plant outside of Oslo for the incineration of organic hazardous wastes. A cement kiln is like a large rotary kiln incinerator, operating at temperatures over 1,500°C. Organic hazardous wastes are fed into the kiln blended with fuel, replacing some of the fuel required to produce cement.

Sweden

Bengt Aplander
National Environment Protection Board
Fack
S–17120 Solna 1
Sweden

West Germany

The West German state of Bavaria, which is about the size of Ohio, has three regional hazardous waste treatment facilities with a combined disposal capacity of 350,000 tons per year. Two of the plants, with a network of collection stations, are run by GSB, a joint venture of the Bavarian state government, three municipal organizations, and Bavarian industry. The third plant is operated by a regional waste disposal company. Both firms are nonprofit. The newest and largest facility is near Ebenhausen. It can handle 160,000 tons per year of hazardous wastes and includes a laboratory, a treatment plant for inorganics, a water purification plant, and the country's largest incinerator. Steam from the incinerator is used for heating, generating electricity to meet the plant's needs, and for selling to the public grid.

Berndt Wolbeck
Assistant Head
Section for Waste Management
Bundesministerium des Innern
Rheindorferstrasse 198
D–53 Bonn 7
West Germany
0288 681–5376

Additional Contacts

Pierre Lieben
Principal Administrator
Environment Directorate
OECD
2. Rue Andre Pascal
75775 Paris Cedex 16
France
502.12.20

Steve Fitt
Department of the Environment
Room 46.09
Romney House
43 Marsham Street
London, 13 PY
United Kingdom
01–228789

NOTES

1. Dutch Parliament, *Act of February 11, 1976, Containing Regulations Relating to Chemical Waste and Used Oil*, commonly known as the Chemical Waste Act of 1976.
2. Franklin Associates, Ltd., "Feasibility Study for Establishing a Waste Exchange in the Rocky Mountain Region," for U.S. EPA Region VIII, report filed in December 1979, p. 10.
3. Report by Vereniging van de Nederlands Chemische Industrie, 1978; assessed by Sandra E. Jerabek for the German Marshall Fund and the Institute for European Environmental Policy in March 1981.
4. For information on America's recycling efforts, contact the Northeast Industrial Waste Exchange, 700 E. Water Street, Room 711, Syracuse, New York 13210. They can supply annual reports of their exchange and current catalogues of their efforts, as well as a May 1983 *Directory of Hazardous Waste Officials in the Northeast* (compiled by Clark-McGlennon Associates of Boston).
5. Frits van Veen, "Treatment of Chemical Wastes," presented at NATO Committee on Challenges to Modern Society meeting on hazardous waste, September 21, 1979. For a full account of these pivotal proceedings, see the following four detailed reports:

1. NATO-CCMS pilot study on disposal of hazardous wastes: chemical, physical, and biological treatment of hazardous wastes in NATO countries. Project

leader: United States. Final Report, NATO Committee on the Challenges of Modern Society, Brussels, 9 March 1981. (145 pages, refs.)

2. NATO-CCMS pilot study on disposal of hazardous wastes: thermal treatment. Project leader: Federal Republic of Germany. Final Report, NATO Committee on the Challenges of Modern Society, Brussels, 23 March 1981. (183 pages, refs.)
3. NATO-CCMS pilot study on disposal of hazardous wastes: metal finishing wastes. Project leader: France. Final Report, NATO Committee on the Challenges of Modern Society, Brussels, March 1981. (281 pages, refs.)
4. NATO-CCMS pilot study on disposal of hazardous wastes: landfill research. Project leader: Canada. Final Report, NATO Committee on the Challenges of Modern Society, Brussels, 16 February 1981. (56 pages.)

These NATO-CCMS pilot studies, only four of a series, should serve to introduce researchers to the global potential for alternatives to land disposal. Copies available from libraries receiving NATO publications; also abstracted on NASA-SCAN data base.

6. Dutch Parliament, *Chemical Waste Act*, Section 34, pp. 11–12.

7. I wish to acknowledge the unpublished papers of Sandra Jerabek and John E. McClure in this area. Since both of these individuals have assessed European facilities and practices firsthand, I have relied heavily on their papers throughout this summary. Mr. McClure wrote his report on "Selected Waste Management Practices" for the Dames and Moore consulting firm. Ms. Jerabek submitted her "An American Appraisal of Hazardous Waste Management in Europe" to the German Marshall Fund in March 1981.

8. For a diagrammatic sketch of the Kommunekemi facility, see Toxic Waste Assessment Group, *Alternatives for the Land Disposal of Hazardous Waste for California* (California Printing Office, Sacramento, Calif., 1982), p. 153.

9. Report by Gary A. Davis; personal communication resulting from work at International Institute for Applied Systems Analysis.

10. For a detailed account of how Kommunekemi is managed, see John E. McClure's report to Dames and Moore, Inc. For copy, contact editor.

11. *NATO-CCMS Pilot Study on Disposal of Hazardous Wastes.* Project leader: United States. Final Report: 9 March 1981. The figure cited is derived from the waste generation volume charts and divided by population, presented in first section of NATO report.

12. For contemporary details, contact Dr. Berndt Wolbek, Assistant Head, Section for Waste Management, Bundesministerium des Innern, Rheindor-Ferstrasse 198, D–53 Bonn 7, 0288 681–5376.

13. Federal Republic of Germany, *Promulgation of the Waste Disposal Act*, as amended January 5, 1977, p. 14.

14. Sandra Jerabek, "An American Appraisal of Hazardous Waste Management in Europe," pp. 7–8.

15. Imperial-Krauss-Maffei Industri-Anlogen Gmblt, *Report: Incineration Plant for Industrial Waste in Bavaria* (Munich, Bavaria, Federal Republic of Germany), p. 14.

16. Gunter Eichele was interviewed by Sandra Jerabek on October 10, 1980

in Bavaria. See "An American Appraisal of Hazardous Waste Management in Europe," pp. 3–4.

17. Ibid., pp. 7–8.

18. For more information on French facilities, see the section on Sidibex in McClure's Dames and Moore report. Contact editor for copy.

19. For more information on the Norcem Cement Plant, and the feasibility of retrofitting American cement kilns for high-temperature toxic waste destruction, contact Harry M. Freeman at California's Department of Health Services at (916) 322–4990 in Sacramento. Also, see Freeman's "Review and Evaluation of Selected Innovative Thermal Hazardous Waste Destruction Processes," written under cooperative agreement no. R–808908 with the Industrial Environmental Research Lab of the U.S. EPA in Cincinnati, Ohio.

Part III
NEW GROUNDS FOR HOPE

9

Preparing Management: The Political Economy of Stronger Controls

PATRICK G. McCANN

The complexity, controversy, potential costs, liability concerns, and business implications of Resource Conservation and Recovery Act (RCRA) requirements have demanded unprecedented corporate management attention. Although comprehensive programs have been implemented at many companies, this is not the time for retrenchment. New and more stringent regulations will have substantial adverse impacts on unprepared firms.

Hazardous waste managers must evaluate the complete range of economic and non-economic factors in charting their course. Economic factors include the trade-offs between on-site and off-site management, as well as the important role of transportation. Non-economic factors to consider include liability protection, service, reliability, and public image.

Managers must not, however, consider these factors only in the current context. The growing public concern that current requirements are inadequate will result in a new wave of hazardous waste regulations. Anticipation and planning must be part of management's thinking if it is to most effectively manage waste in a rapidly changing regulatory environment. To keep on top of the situation, companies should take an active role in the rule-making process, develop detailed profiles of their current management approaches, and develop contingency plans for alternate regulatory scenarios.

More stringent regulation may worsen capacity shortfalls that exist in certain regions of the country. To stimulate the development of environmentally sound treatment and disposal, states and the federal government must take positive action. Policymakers have two approaches to the problem: either from the demand side by changing the cost to generators of treatment and disposal, or from the supply side by creating new capacity directly through state ownership or other institutional mechanisms.

The next five years will challenge corporate hazardous waste managers. Success will depend on anticipation, careful planning, and effective implementation.

INTRODUCTION

Director of manufacturing, director of marketing, director of corporate development, vice president of engineering, plant manager. Are these attendees at a company sales meeting? Perhaps, but also some of the key company personnel at a chemical firm's recent meeting on hazardous waste issues.

"Cradle to grave" regulations implementing the Resource Conservation and Recovery Act have affected a broad range of corporate decisions from the type of material used to paint auto and truck bodies to the marketability of a new chemical solvent that is more easily recoverable.

RCRA requirements have demanded unprecedented management attention throughout business organizations. Beginning in the late 1970s, leading companies developed comprehensive, three-part programs to deal with their wastes and the new regulatory climate. The programs involved:

- *Strategy*. Corporate managers, once made aware of the scope of their problem, initiated studies of waste generation and of current treatment and disposal practices. As a result, corporate policies concerning treatment and disposal options and cost/liability trade-offs were developed.
- *Tactics*. Responsibility for hazardous waste management was centralized in some companies; in others, plant personnel retained authority but approval of waste management plans by corporate staff was required. Policies and programs were standardized, and compliance audit programs were initiated.
- *Operations*. Implementation of corporate planning to institutionalize hazardous waste management was the final step. As the result of such attention, companies often found that they could reduce waste generation by 10 to 30 percent.

Development of such across-the-board programs seems to imply that corporate hazardous waste management is now well in hand, and that management can now shift its attention to more pressing concerns. This would be a mistake. Since increasing pressure will force hazardous waste generators to improve the management of their wastes, companies unprepared for such pressure will be at a substantial disadvantage to their counterparts. Anticipating the pressure is a key to positive action.

In this chapter, the economics of hazardous waste management will be examined. Although the primary subject is the manager responsible for hazardous waste in the waste-generating firm, the implications of eco-

Figure 1
Overview of Decisionmaking Process

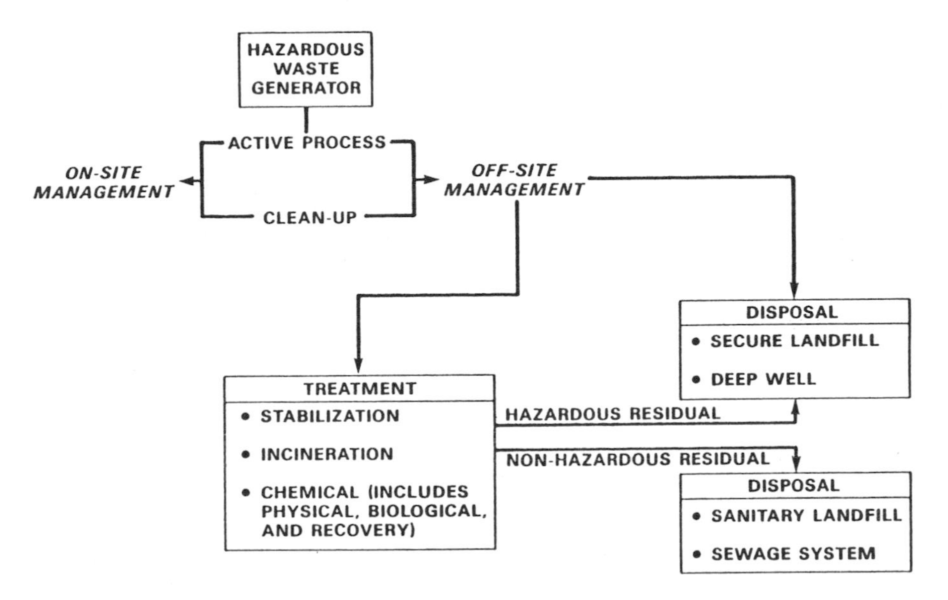

nomic decision making to policymakers will also be highlighted. The chapter is organized into three sections.

"The Decision-Making Process" explores the options available for hazardous waste treatment and disposal and presents some typical costs for on-site and off-site management. The importance of non-economic factors such as liability protection, reliability, and service are also explained. "Planning for the Future" examines the need for continued top management attention and suggests ways to anticipate and plan for future changes in waste management requirements. Finally, "Government Incentives to Stimulate New Capacity" presents a number of economic incentives available to policymakers to stimulate new capacity.

Meant to serve as a foundation in the economics of hazardous waste management, this chapter also recognizes that individual generators face widely different options depending on the type of waste generated, state regulatory requirements, geographic location, and their ability to use waste by-products in the production operation.

THE DECISION-MAKING PROCESS

It is useful first to understand the basic options available to hazardous waste generators. Figure 1 presents a flow diagram of the decisions facing a generator. The first decision is whether to manage wastes on site or ship them off site. A generator may then have to choose which treatment and disposal steps are appropriate for a particular waste stream. Some wastes may be rendered non-hazardous through treatment; others may be subjected to treatment for waste volume reduction; and still others may go directly to disposal. Waste management options may be divided according to six general categories: treatment, resource recovery, incineration, deep well injection, secure landfill, and land treatment/solar evaporation.

Treatment includes various forms of chemical treatment such as neutralization of acids and caustics; physical treatment such as oil skimming and sedimentation; and aerobic and anaerobic biological treatment. Treatment often results in a byproduct sludge that must be disposed. This sludge may or may not be hazardous and may be about 10 percent of the original volume treated.

Resource recovery often involves processes similar to chemical treatment, which differ only in that the waste is partially or totally transformed into a product for sale. By far, solvent recovery is the most widely reclaimed "waste." Resource recovery processes also may result in a byproduct sludge that must be disposed.

Incineration is the controlled burning of combustible solid, liquid, or gaseous wastes. Incineration results in by-products that may or may not be hazardous. Incinerator ash may contain metals that would make the residues hazardous. Similarly, the waste sludge from the stack scrubber may contain hazardous components. Although incineration results in these by-products, the volume of waste can often be drastically reduced and the degree of hazard associated with the waste substantially reduced.

Deep well injection entails pumping liquid waste into porous underground formations isolated from potable-water-bearing formations. Deep well injection of hazardous waste requires special geologic conditions not present in much of the country.

Secure landfill involves burial of drummed or bulk materials consistent with standards specified in federal and state regulations. Recent federal regulations now require synthetic liners and leachate collection systems on new landfills.

Land treatment/solar evaporation is a form of biological treatment through soil incorporation. Organic wastes are applied onto or beneath the soil and then periodically mixed to aid in the decomposition of the organic material. Often land treatment starts with solar evaporation of the aqueous portion of the waste stream.

Although there are many treatment and disposal options, each option is appropriate only for certain types of waste streams. For example, inorganic acid waste is not suitable for incineration, nor is an organic still bottom solid suitable for deep well injection. Figure 2 compares the treat-

Figure 2
Waste Treatment and Disposal Options Versus Waste Type

● OFTEN USED
◒ SOMETIMES USED

		TREATMENT/RECOVERY				DISPOSAL				
		CHEMICAL	PHYSICAL	BIOLOGICAL	WASTE EXCHANGE	LANDFILL	LAND FARMING	DEEP-WELL INJECTION	LIQUID INCINERATOR	ROTARY KILN INCINERATOR
ORGANICS	OIL AND OILY SLUDGE		●	●	●		●	◒	●	●
	PAINT AND PIGMENT RESIDUE		◒			●				◒
	PLASTICIZERS					●				●
	SOLVENTS		●		●				●	●
	WASTE WATER TREATMENT SLUDGES	◒	◒	◒		●				
	PESTICIDES	◒		◒		●			●	●
	STILL BOTTOMS		◒			●				●
	PHARMACEUTICAL WASTE			◒		●			●	●
	OTHER	◒	◒	◒	◒	◒	◒	◒	◒	◒
INORGANICS	ACIDS	●			●			●		
	ALKALINE	●			●			●		
	CYANIDE	●						●		
	HEAVY METALS		●			●				
	WASTE WATER TREATMENT SLUDGES		●			●				
	CATALYSTS				◒	●				
	OTHER	◒	◒	◒	◒	◒	◒	◒	◒	◒

ment and disposal options most often used for major categories of hazardous waste.

Although most industries generate some hazardous waste, the chemical, primary metals, and the fabricated metal products industries account for over three-fourths of the hazardous waste generated in the U.S.[1] In addition to variety in the kinds of waste generated, industries may have different options available to them as a result of the following factors:

- *Size.* Smaller generators may not be able to realize the economies of scale needed for on-site treatment and disposal and may be forced to rely on off-site management.
- *Uniformity of waste streams.* Generators producing large volumes of a single waste stream are better able to custom design a treatment or recovery system than a generator with a large number of intermittent streams.

- *Use of by-products.* Some generators can use wastes in other portions of their production process, for example, as a fuel supplement in a boiler or an acid in a chemical process.
- *Location.* Geologic conditions, climate, land availability, and proximity to commercial treatment and disposal facilities all have an important influence on the options available to generators.

Although these factors certainly make hazardous waste management a highly site specific process, there are a number of economic and non-economic factors that should be considered in any situation. Often at the top of the list is liability control. Many have learned the hard way the meaning of liability from handling waste cases. This concern tends to make firms wary of land disposal and more likely to manage wastes themselves rather than rely on commercial off-site facilities. For example, one of the largest chemical firms currently manages over 95 percent of its own wastes and will not land dispose any waste that can be reasonably incinerated.

Generators also cite serviceability and reliability as key factors in choosing among waste management options. Should one risk production interruptions although it might involve disproportionately large costs to the firm to do otherwise? If off-site facilities are used, can the hauler or disposal firm respond on short notice and accept all wastes the firm generates? Will the approach taken allow future process changes or expansions? Generators should also consider how adverse public reaction could focus on the firm in the event it experiences real or perceived problems with hazardous waste management. For example, a baby powder producer may be substantially more sensitive to a mention of a problem with its waste management practices in the press than a steam turbine manufacturer.

These "non-economic" factors are certainly important, but generators do not ignore the costs of management. In fact, aside from potential liability considerations, cost is often the most important factor cited by waste managers.

The cost of on-site or off-site management includes all storage, handling, transportation, treatment, and disposal costs including labor, materials, and overhead. Table 1 presents capital and operating costs for new hazardous waste management facilities. Although these costs vary considerably, it is clear that a full scale treatment or disposal facility requires a substantial investment. For many generators, on-site management is less expensive than the cost of shipping wastes elsewhere.

Off-site prices charged for treatment and disposal more than doubled from 1976 to 1980. In a few cases, landfills that previously charged $5 to $10 per ton increased prices to over $50 per ton. These price hikes reflect the more stringent regulatory requirements as well as the shortage of

Table 1
Typical Costs for New Hazardous Waste Management Facilities

Type of Facility	Throughput	Capital Costs* (Millions)	Annual Operating Costs (Millions)
Chemical Treatment	50,000 tons/year	$2	$2
Rotary Kiln Incinerator	25,000 tons/year	$15	$5
Liquids Incineration	5,000 tons/year	$5	$1
Secure Landfill	35,000 tons/year (20-year life)	$4	$2

*Excludes facility permitting and site purchase, which is highly variable. Some firms quote estimates of $1 million to $3 million.

available capacity. Table 2 provides a comparison of quoted prices for off-site management in 1981.[2] Recently, the price explosion has leveled off. Prices for commercial waste management have changed very little since 1981. Intense competition among off-site firms and reduced demand due to the recession have kept prices at roughly their 1981 levels.

Prices for off-site service vary considerably, depending mostly on the level of regional competition and the type of waste. The outlook for future prices also varies considerably.

Landfilling constitutes about one third of the revenues of the hazardous waste management industry. Although landfill prices have remained fairly stable over the last two years, there is considerable regional variation. Landfill prices are usually based on whether the material is received in drums or bulk and whether any solidification of liquids is needed prior to land disposal. Drum prices are usually higher on a per pound basis. Solidification[3] can add $10 or more per drum to the disposal costs depending on the amount of handling required and the volume increase after bulking materials are added.

Incineration prices vary to a greater extent than any other treatment or disposal service. Prices generally depend on shipping mode (drum vs. bulk); physical state (solid, sludge, or liquid); degree of halogenation; BTU content; and ash content. Although prices do vary considerably, they generally fall into three categories. *High BTU waste liquids*, such as spent solvents, are accepted for very low prices, and in some cases, where the waste can be used as a fuel supplement in cement kilns or light aggregate manufacturing, generators have been paid $.05 or more per gallon for their "waste." *Low BTU wastes* that cannot be used as a fuel generally command a higher price because commercial incinerators require a fuel supplement to achieve the required destruction efficiency.

Table 2
Comparison of Quoted Prices for Off-Site Hazardous Waste Management Services for Nine Major Waste Firms in 1981

Type of Waste Management	Type or Form of Waste	Price	$/Ton
Landfill	Drummed	$0.64-$0.91/gal ($35-$50/55 gal drum)	$168-$240
	Bulk	$0.19-$0.28/gal	$55-$83
Land treatment	All	$0.02-$0.09/gal	$5-$24
Incineration clean	Relatively clean liquids high-Btu value	$(0.05)-$0.20/gal	$(13)-$53
	Liquids	$0.20-$0.90/gal	$53-$237
	Solids, highly toxic liquids	$1.50-$3.00/gal	$395-$791
Chemical treatment	Acids/alkalines	$0.08-$0.35/gal	$21-$92
	Cyanides, heavy metals, highly toxic waste	$0.25-$3.00/gal	$66-$791
Resource recovery	All	$0.25-$1.00/gal	$66-$264
Deep well injection	Oily wastewater	$0.06-$0.15/gal	$16-$40
	Toxic rinse water	$0.50-$1.00/gal	$132-$264
Transportation		$0.15/ton mile	

Source: Booz-Allen & Hamilton, Inc.

Prices charged for these kinds of wastes generally range from $.20 to $.90 per gallon. *Organic solids and sludges* command the highest incineration price due to handling problems. These wastes are typically shipped in drums and very few commercial incinerators currently have the ability to accept drums. Prices range from $1.50 to $3.00 per gallon for these wastes.

Chemical treatment and resource recovery prices, as one would expect, also vary tremendously. Prices depend on the complexity of the treatment and the number of unit processes required. Aqueous treatment of dilute wastes may cost as little as $.08 to $.10 per gallon. Simple neutralization of acids or caustics usually costs $.20 to $.30 per gallon. More sophisticated treatment of more complex wastes, such as cyanides or highly toxic or reactive wastes, may cost more than $2.00 per gallon.

Deep well injection is a low-cost option for disposing of a number of pumpable liquid waste streams. It is used widely for disposal of aqueous organic and inorganic wastes in the Gulf Coast. Due to slack demand, deep well prices have remained fairly constant for a number of years averaging $.05 to $.15 per gallon.

Land treatment of wastes is a common technique used in the Gulf and West Coast for disposing of petroleum-related wastes. Due to the high net evaporation rate in these areas, this can be a very inexpensive method of waste management with prices ranging from $.02 to $.10 per gallon.

Transportation costs are often the deciding factor for many firms in comparing the costs of on-site versus off-site management. The cost of shipping waste to a commercial facility can be 50 percent or more of the total disposal cost. Transportation costs depend on the need for specialized equipment and the distance involved: a practical rule of thumb is $.10 to $.15 per ton per loaded mile. For example, a shipment of 80 drums each weighing 500 pounds transported 200 miles could cost about $400 to $600 (20 tons × 200 miles × .10 to .15 per ton mile) or $5.00 to $7.50 per drum.

The ultimate choice of on-site versus off-site management is a complex decision involving economic as well as non-economic factors. Yet in general, generators of large volumes of easily treatable wastes tend to choose on-site management, while firms with small volumes of relatively hazardous waste tend to opt for off-site management.

PLANNING FOR THE FUTURE

The rising strength of public opinion against land disposal is the primary force behind the new and more stringent hazardous waste regulation. Every new discovery of a hazardous waste problem only reinforces the public's belief that current requirements are inadequate. Despite all that management now does to control waste generation, treatment, and disposal, it is dogged by the past. The recent political controversy over the EPA and the administration's commitment to a strong hazardous waste policy further reinforces the public's concern. Another wave of stronger controls is imminent.

Corporate managers who doubt that these regulations will have substantial impacts on their operations need only look at the last five years of regulation and ask themselves if past regulations had significant impacts. The exact form of the next wave of hazardous waste regulations is difficult to predict, but a number of developments are generally accepted as likely.

Land disposal ban on high profile wastes. The first step toward more stringent regulations probably will be a national ban, imposed as early as 1984, on land disposal of certain high profile, politically sensitive wastes. These include PCBs, cyanides, dioxins, and possibly some pesticides. EPA currently is considering criteria, including extreme persistence, mobility, and toxicity, to identify and list wastes unfit for land disposal. Commercial waste management firms, however, object to such listings because very little of these wastes are disposed of at landfills. A formal ban, they argue, will make acceptance of any wastes at any facility more difficult.

Land disposal ban on burnable or treatable wastes. A number of states already have taken steps to ban land disposal of wastes that can be burned or treated by other methods. For example, Missouri now prohibits wastes with organic contents greater than 5 percent from landfill disposal. New York has considered banning all organics from landfills, and still others like California are taking a phased approach to banning land disposal of highly toxic but treatable wastes.

Increased enforcement. Another major regulatory development will be increased EPA enforcement action. Such action will take place in part as a response by the Reagan administration to its enforcement critics. Once visible actions are taken under the EPA's revamped enforcement organization, pressure will build for manifesting clear results and improvements. Such pressure may also be translated into a campaign issue in 1984.

Tightening of small generator exemptions. Those advocating exemptions for small volume waste generators often contend that economic waste management alternatives are not available to these firms and that the government's limited regulatory resources should be focused elsewhere. However, this viewpoint is unlikely to prevail much longer, especially against arguments that indicate the exemptions exclude 25 percent to 30 percent of all hazardous waste generators from regulation. It is likely the EPA will tighten small generator exemptions soon, which will have a significant impact on a number of industries. For example, pesticide formulators and dealers may have to change distribution strategies to help customers meet new requirements. Farmers or pesticide applicators may be reluctant to use certain compounds whose use will require greater administrative paperwork and costs.

Toxic victim compensation fund. Momentum is building on Capitol Hill to develop legislation covering personal liability damages. Although funding such a program would be as controversial as funding Superfund, hazardous waste generators and commercial waste management companies are likely to be substantial contributors to the fund.

With these changes in the wind, hazardous waste managers should realize that policymakers will continue to develop their approaches to regulation in response to scientific data and public concern for adequate protection. We are in a learning curve, and no "final solution" is yet on the horizon. For this reason, three steps are recommended:

- *Take an active role in rule making.* Regulatory managers must continue to be vigilant in monitoring and taking part in the rule-making process. Providing information early in the process often is the best opportunity to ensure that rule making reflects the impact on all generators.

- *Develop waste generation profiles and consider disposal and treatment alternatives.* Companies should not be content simply to comply with existing requirements. Instead, waste streams and disposal and treatment methods should be carefully evaluated in terms of future regulation. The goal is to minimize new risks and waste management costs. By identifying wastes likely to be banned from landfills or requiring new treatment methods, potential problems can be avoided in a timely and efficient manner.
- *Develop contingency plans.* Hazardous waste managers should investigate the feasibility and cost of raw materials substitution or alternatives that could eliminate or minimize disposal and treatment problems. Future cost and liability issues are of key importance. For example, if a particular facility relies on on-site landfill disposal of an organic sludge that may be banned from such disposal, options such as substitution, installation of an incinerator, or off-site contracting for disposal should be considered.

During the next several years, both federal and state regulators will be implementing new and more stringent requirements to control hazardous wastes, especially with respect to land disposal. Their actions will comprise the next wave of hazardous waste regulation, and will come in response to unabating public pressure for more adequate health and environmental protection.

Consequently, corporate compliance managers who anticipate, plan, and act on these regulatory changes will be at a substantial advantage over those who are retrenching. The mere fact that policymakers view regulation as an evolving process (see chapter 2) requires corporate managers to regard their hazardous waste strategies in the same manner.

GOVERNMENT INCENTIVES TO STIMULATE NEW CAPACITY

More stringent regulations will decrease the risks posed by inadequate waste management practices. Unfortunately, these regulations may also temporarily narrow the range of options available to generators for waste management. For example, the January 26, 1983, land disposal regulations, which called for synthetic liners and leachate collection systems, caused a number of commercial landfills to close. This worsened capacity shortfalls in certain regions of the country.

To counteract the impact of reduced capacity to treat and dispose of hazardous waste, almost every state, as well as the federal government, has considered how to stimulate the development of new capacity to treat and dispose of hazardous waste in an environmentally safe manner. In a nutshell, the "hazardous waste capacity" problem involves (1) Intense public opposition to siting new facilities using existing technology, which makes additions to capacity almost impossible; and (2) Insufficient demand for more sophisticated treatment and disposal technologies to justify the addition of these services at existing facilities and competition from the more traditional, less expensive methods, which make these options not cost competitive.

Policymakers can approach this problem in two ways: from the demand side by changing the cost to generators of various treatment and disposal technologies; or from the supply side by creating new capacity directly or by providing assistance overcoming some of the institutional hurdles to develop new capacity. Many states have opted for the demand-side approach. Three variations have been used:

- Promote facility development through legislation creating siting boards or by designing institutional mechanisms for overcoming public opposition to siting.
- Select and purchase a site and lease it to a private company for development and operation.
- Select and purchase a site and then utilize the facility as a state-run operation.[4]

Each option has advantages and disadvantages. A number of states have chosen to own the site but offer a contract for its operation. This option provides an assurance to the public of long-term care, alleviates the siting problem, and yet relies on private enterprise to assume the business risks of operating the facility.

An alternative is to stimulate capacity from the demand side by changing the relative cost of new treatment and disposal options. By discouraging less desirable forms of treatment or disposal, demand will be shifted to environmentally safer but more expensive options. For example, some states have enacted taxes on land disposal of wastes. However, the price discrepancy between land disposal and alternative treatment or disposal options is usually much greater than the tax, so it has had little effect on the choice among waste management options. Furthermore, if similar taxes do not exist in neighboring states, generators may simply opt to landfill wastes across the state line. A national strategy, therefore, would be needed to improve treatment and disposal from the demand side.

NOTES

1. *Hazardous Waste Generation and Commercial Hazardous Waste Management Capacity—An Assessment*, Booz-Allen & Hamilton for the U.S. EPA, November 1980.

2. *Review of Activities of Major Firms in the Commercial Hazardous Waste Management Industry: 1981 Update*, Booz-Allen & Hamilton for the U.S. EPA, May 1982.

3. Current federal requirements prohibit drummed liquids to be disposed in landfills; therefore, solidification of liquids is required prior to landfill disposal.

4. *Comprehensive Hazardous Waste Management Facility Study*, prepared for the Georgia Hazardous Waste Management Authority, November 1981.

10

Emerging Options in Waste Reduction and Treatment: A Market Incentive Approach

JOEL S. HIRSCHHORN

Public policies protecting Americans from the threat of hazardous wastes can be achieved by two strategies. One is the regulatory approach and the other is the market or economic approach. The federal government's regulatory approach is based on two primary statutes: the Resource Conservation and Recovery Act (RCRA), which governs the management of hazardous wastes currently produced; and the Comprehensive Environmental Response, Compensation, and Liability Act (CERCLA). Better known as the Superfund, CERCLA governs the management of accidental spills of hazardous substances and also finances the emergency cleanup of releases of hazardous substances from uncontrolled waste sites.

Although both RCRA and CERCLA indirectly influence the costs and liabilities of waste management, their economic effects are understood to be uncertain. This is mostly because the indirect economic effects of these regulations are themselves victimized by the slow, ever changing, and incomplete nature of the federal regulatory waste program. Other complicating factors include failures and delays in the state regulatory programs meant to implement and supplement RCRA and CERCLA.

As yet there are no direct federal market incentives to achieve improved private sector management of hazardous wastes. However, as discussed below, some states have taken a stronger stance by employing direct economic pressures.

The time has come to seriously consider direct market incentives at a federal level in order to achieve a number of desired protection goals. This chapter presents the case for implementing these new tools in the

The views expressed here are not necessarily those of the Office of Technology Assessment (OTA).

following fashion. First, a brief review of current practices and problems serves to establish the need for direct market incentives. Second, a basic approach is defined that could introduce direct market incentives within the present federal program. Third, the relationship of the proposed approach to state programs is elaborated including a discussion of several state programs that illustrate the feasibility of a federal approach. Fourth, the likely objections to the proposed approach and the institutional issues left for resolution are examined. Finally, a summary situates the proposal in the context of other ongoing efforts to improve and expand the federal hazardous waste program.

THE NEED FOR DIRECT MARKET INCENTIVES: CURRENT PRACTICES AND PROBLEMS

As a recent congressional report[1] on the national hazardous waste problem emphasized, land disposal continues to be the dominant management choice, accounting for as much as 80 percent of all hazardous wastes. Furthermore, as OTA and a number of other studies have concluded, land disposal of hazardous wastes cannot—even under the new federal environmental regulations set by the Environmental Protection Agency (EPA)—ensure adequate protection of public health and the environment, either in the near term or decades from now. Current EPA regulations do not place stringent enough requirements on land disposal facilities to either prevent, detect, or correct releases of hazardous waste constituents into the environment (see chapter 1).

In addition to the technical inadequacies of these regulations, another undesirable effect is that land disposal continues to be cheaper than employing safer alternatives. The lack of regulatory controls against inexpert dumping contributes to artificially lower prices for land disposal. Although it is impossible to prove quantitatively, most experts believe that if the government regulations for land disposal were better designed as well as more vigorously administered and enforced, then the market prices for land disposal would rise substantially. The costs of ensuring total isolation of wastes from the environment, of maintaining effective monitoring, and of providing capabilities for corrective action are believed to be so great as to make land disposal options far less competitive than they currently appear. The net result is that those who use land disposal, either on site for waste generators or off site at commercial facilities, do not pay the real, full, long-term costs for their services. Instead, much of the cost is externalized or shifted to government or to society as a whole. This incomplete internalization of the costs of land disposal constitutes a market distortion and economic inequity. This is what keeps land disposal the dominant mode of waste management.

Moreover, it is now clear that for many wastes land disposal is inappropriate and unnecessary. Land disposal should be used only for certain types of wastes and only in select locations. For some time, the choices facing managers of hazardous wastes have been seen as a hierarchy. The most desirable options are those that can reduce the amount of hazardous waste at the source, such as separation of wastes, recovery and recycling, process changes, and end-product substitutions. The second favored option is any type of physical, chemical, biological, or thermal treatment that permanently reduces or eliminates the hazardous characteristics of the waste. The least desirable waste management practice is land disposal.

Moreover, for those wastes that are highly toxic, persistent, mobile, and bioaccumulative, land disposal is clearly unacceptable, even if it were better regulated. However, it must also be recognized that there are hazardous wastes, such as residues from treatment operations, for which land disposal—properly regulated, adequately designed, and effectively operated—is technically and economically appropriate.

The most important news on the hazardous waste problem is that there are technically feasible alternatives to land disposal for virtually all hazardous wastes. As a result of this technical documentation, the primary issue today is: *Why are the technological alternatives to land disposal of hazardous waste not being used to a greater extent?*

The chief answer is that there is currently insufficient economic motivation to use these alternatives. In most cases, land disposal costs less than the alternatives, typically from $50 to several hundred dollars less per ton of waste. There is, of course, no substitute for having effective regulations to ensure that everyone is fulfilling the same standards of safety. However, as a complement to the regulatory approach, which can have some limited effects on raising the costs of land disposal, it may be necessary to introduce direct economic incentives for the greater use of alternatives to land disposal. Once this is accomplished, other aspects of the problem can be more easily resolved, such as overcoming regional shortages of treatment facilities, as well as conquering difficulties with siting. With the introduction of a federal incentive program for alternatives to land disposal, the general public will be better educated about the differences between land facilities, which have a poor history of performance, and treatment facilities, which are often no different than conventional manufacturing operations.

Another factor favoring alternatives is that the costs of cleaning up uncontrolled sites under the Superfund program are becoming larger. The OTA study estimated that to clean up America's 15,000 uncontrolled sites would likely cost at least $10 to $40 billion. For some of the worst sites, cleanup may not even be feasible, and for others the long-term effectiveness of cleanup actions are not yet determined. Cleaning up land disposal sites (which constitute most uncontrolled sites) will generally cost 10 to 100 times as much as it would have cost initially to manage the wastes so as to avoid future cleanups. The challenge, therefore, is to stop creating future uncontrolled sites. In addition to having more stringent and effective land disposal regulations, the answer is to stop using land disposal options for most hazardous wastes.

A FEDERAL FEE SYSTEM FOR HAZARDOUS WASTE GENERATORS

The OTA study considered the benefits of introducing a federal fee system on hazardous waste generators. This would serve the dual purpose of raising revenues to support an expanded Superfund program while it influenced current waste management decisions. By using a structured system of different fees for a relatively small number of waste types and management choices, it would be possible to shift beyond dumping to the more desirable categories of waste reduction and treatment. No new federal program would have to be introduced.

The key of the system would be to change the current mechanism of financing the Superfund program from fees on chemical and petroleum feedstocks to fees on hazardous waste generators. Using feedstocks as the basis of financing Superfund made sense in 1980 when there was little information on hazardous waste generators and their wastes. But now better information is available and virtually everyone agrees that there is little connection between feedstocks and the quantity and types of hazardous wastes ultimately produced, nor to the quality of their management. Moreover, the feedstock fees penalize those companies producing the least hazardous waste and those that use alternatives to land disposal. This is because they too must pay the same fees as their competitors who contribute more hazardous waste. By switching from feedstocks to wastes, CERCLA and RCRA can become better integrated into one federal program. Clearly, greater equity would be achieved by shifting the burden of supporting Superfund from a relatively small number of companies to a larger number of waste generators and to those consumers who buy the products and services that contribute to hazardous waste production.

The federal waste fee system would consist of the following points, meant to shift management choices away from land disposal:

1. Fees should be much higher for most hazardous wastes currently disposed of on or into the land.
2. Fees should be lower for wastes that are chemically, thermally, biologically, or physically treated.
3. Fees should be lower for high volume–low hazard wastes, such as those deemed hazardous because of characteristics other than toxicity.

4. There should be no fees for wastes that are recovered or recycled on site for their energy or material value.
5. There should be higher fees, or a supplemental fee, on wastes transported to off-site or commercial facilities because of the potential for accidents and spills during transportation.
6. There probably should be higher fees on wastes disposed of in old land disposal facilities, since they are regulated less stringently than new ones.

An important general rule for such a system is that it should be simple, using several generic categories of wastes which reflect fundamentally different hazard levels and only a few categories for waste management. This would facilitate administration on the basis of readily available information.

Fees must be high enough to accomplish two purposes: to affect current decisions concerning the use of land disposal or its alternatives, and to raise sufficient funds for Superfund activities. To affect private sector decisions, it will be necessary to have fees that, in some cases, will add from $50 to perhaps $100 and more per ton. To raise more than the original $1.6 billion planned for the five-year period of Superfund through 1985, it will be necessary to have fees that average more than $10 per ton. Although the desired amount of money to be collected under Superfund is uncertain, it is becoming increasingly clear that it may be necessary to raise perhaps two to three times the annual amount in the current program. Moreover, Superfund monies also might be used to support state hazardous waste regulatory programs and a victim compensation program excluded from the initial Superfund program.

RELATIONSHIP TO STATE PROGRAMS

A number of states have already taken action in the direction of this proposal. Spurred by money problems, states have been pursuing a variety of fee approaches to raise funds for state Superfunds or for ongoing hazardous waste regulatory activities. In some cases, there has also been interest in affecting current management choices (see chapters 6, 7, and 8). It should be noted, however, that in most cases to date the levels of fees confronting waste generators are too low to affect most decisions based primarily on economics. Since pressures within a state can be severe, including the threat of manufacturing operations moving elsewhere and workers facing job loss, these state programs often have been forced into levying relatively low fees.

There are other limitations that make it difficult to rely on state fee systems to shift management decisions away from land disposal. Many states cannot legally impose such fees, and others will choose to avoid them. This raises the spectre of forming "pollution havens" when other states impose reasonably high fees. Moreover, for those states that do impose fees, there are likely to be large differences in the fee structures adopted. Thus, a federal waste fee system is needed to achieve equity, economic consistency, and equal public protection nationwide.

An interesting waste fee system was proposed in Minnesota. There were three management options: land disposal, land disposal after treatment, and treatment. For the last two, rates were doubled for off-site management. For all three categories, rates were doubled when the wastes were liquid rather than solid. Rates ranged from a high of $42 to $85 per metric ton for land disposal to a low of $5 per metric ton for on-site treatment of solid wastes.

New York State employs a simple system, with the following rates imposed on waste generators: $12 per ton for landfilled wastes; $9 per ton for wastes treated or land disposed of off site, excluding landfilling; $2 per ton for waste treated on site. There is no application of the degree-of-hazard concept, the levels of fees are likely to be too low to affect most management choices, and, like the Minnesota system, there are no fees for recycled wastes.

California imposes a fee of $4 per ton for wastes that are land disposed (with a limit of $10,000 per month). In addition, another fee is based on a fluctuating base rate meant to total $10 million annually. In 1982, the base rate was $6.52 per ton for land disposal. This is doubled for extremely hazardous wastes; a rate of 15 percent of the base rate is imposed for wastes put in surface impoundments and for non-federally regulated wastes; while a rate of 0.1 percent of the base rate is imposed for mining overburden wastes. There are no fees on treatment options and there is no distinction made for on-site versus off-site management.

OBJECTIONS TO A FEDERAL WASTE GENERATOR FEE SYSTEM

There will be significant objections to changing the basis of funding CERCLA from fees on feedstocks to fees on hazardous waste generators. Clearly more parties would face payments, more funds may be collected than under the original program, and more specific data concerning waste management would become integrated into federal activities. The purpose of the following discussion is to anticipate these objections and to illustrate how they can be reconciled.

Is it feasible to implement such a federal system? It would have been difficult to implement this approach in 1980 when CERCLA was initiated because of the lack of data on waste generators and wastes. Although far from complete, the data situation is substantially improved today due to various federal and state requirements under RCRA. Introducing a federal waste fee system would also provide considerable incentive to improve and maintain the necessary data base. There are currently about 60,000 hazardous waste generators being regulated, and even if expansion of RCRA by Congress should increase this figure by a factor of two or more, the number of parties from which fees would be collected nationwide is quite manageable. Moreover, as of April 1, 1983, the new CERCLA tax of $2.13 per ton of hazardous waste is being collected to establish the $200 million Post-Closure Liability Trust Fund. This effort further facilitates the proposed system—especially since the present feedstock system will expire in 1985. Finally, administering a federal waste fee system presents far fewer problems as compared to administering technically complex waste regulations. To the extent that the fee system leads to less waste production and more use of alternatives to land disposal, administrative burdens would be reduced.

Should the fee be levied against waste management facility operators rather than waste generators? The basic reason for imposing the fee on generators is that critical decisions concerning waste management, including whether or not to pursue waste reduction efforts, are made by generators. To achieve the maximum effect from a waste fee system on current management practices, it is necessary that waste generators feel the economic effect of the fees. While it is true that fewer parties would be involved if fees were collected from waste management facility operators, it is necessary to point out that about 80 percent of hazardous wastes nationwide are managed on site by generators themselves. Moreover, if commercial off-site facility operators pay the fees, then there is no assurance that they will be completely or uniformly passed on to waste generators. Various market pressures could lead to other practices.

Will a federal waste fee system lead to more illegal dumping of hazardous wastes? Some believe that adding more to the costs of land disposal will cause some generators to dump their wastes illegally. Logically, this means that an added economic cost will cause some generators to break two laws rather than one, with the new one involving the federal Internal Revenue Service. While this might occur in some instances, it must be emphasized that regardless of the imposition of the proposed fee system, it is necessary to have effective and forceful enforcement programs that make any illegal waste management practices highly vulnerable to detection and criminal prosecution. Use of a federal waste fee system should not be viewed as a replacement for normal legal remedies to hazardous waste problems. But the waste fee approach does offer a more direct way to influence the quality of waste management practices.

Will there be problems due to inadequate capacities in treatment facilities, and will there be critical investment and research and development needs? There is evidence that there is considerable unused capacity in commercial treatment facilities; in 1981 about 20 percent for incineration facilities, about 40 percent for chemical treatment ones, and about

75 percent for resource recovery ones.[2] However, there may be regional problems with insufficient capacities at reasonable distances from waste generator locations. Some waste generators and commercial waste managers may need capital for waste reduction or treatment equipment, or they may encounter technical problems requiring research and development. The chief solution to both needs is to tie the federal fee system to federal assistance for capital and research and development needs, in a way that inspires the desired changes in hazardous waste management. This was part of the OTA waste fee system proposal. Assistance can be provided in ways that do not necessitate direct federal outlays, such as by loan guarantees. A portion of the federal fees collected could be used to support these assistance programs.

Will fees distort the marketplace? There are some concerns that waste fees would disrupt the marketplace by, for example, raising prices for some products. Even a marked increase in the cost of land disposal would affect final consumer prices by a few percent at most, since waste management costs currently represent a very small percent of final consumer prices. For example, OTA estimated that all total current waste management spending is about $4 to $5 billion annually. This represents a rather small fraction of the national economy and only about 10 percent of all industry spending related to environmental protection. Moreover, fees on wastes would correct a current market distortion by placing burdens where they belong. Finally, since hazardous waste offers no social benefit, the goal is to create economic feedbacks that prompt the nation to produce less waste.

Will fees place too heavy a burden on industry? Many fees will represent a small fraction of current waste management costs, probably from 5 to 20 percent for many cases. But in other cases, costs may increase from 50 to 100 percent. Naturally, fees must be great enough to affect waste management decisions. Adverse effects on generators can be minimized by (1) providing sufficient time before implementation to allow planning and changes and (2) providing federal assistance for capital and research and development needs. Moreover, if current gaps in RCRA regulatory coverage are closed, more wastes will be regulated, and this can contribute to lower fees. Over time, fees on the least desirable waste management practices can be increased, while fees on other options can be decreased. If the fee system is having the intended effects, then Superfund needs will decrease over time, which will allow lower fees.

SUMMARY DISCUSSION

There is considerable support for the federal waste generator fee system under CERCLA. All those who recognize the inadequacies of the regulatory approach to achieving greater protection of public health and the environment from hazardous wastes will support a direct market approach to complement regulations. Those companies paying an inequitable and disproportionate share of Superfund costs presently will favor, as they did in 1980, the shift away from fees on feedstocks to wastes. Already, bills have been introduced in the U.S. Senate and House of Representatives based on the waste fee proposal. For those who believe that considerably greater sums must be provided under Superfund for cleanups or for victim compensation, shifting to a broader based waste fee system provides greater capability to generate more funds equitably.

The time appears most appropriate for introducing the waste fee system at the federal level and support from the states will be assured if some portion of the funds collected, such as 10 percent, is used to support state hazardous waste programs. This could replace and provide more funds than the current RCRA grant program and perhaps replace the matching fund requirements under CERCLA. Attempts by states to raise more of the required revenues appear ineffective. Thus, the federal waste fee proposal offers a reasonable way to meet state needs in an equitable manner.

Economists and others have long argued for using economic incentives to achieve environmental protection goals. The mechanism of funding CERCLA offers a remarkable opportunity to do just this in an area that poses considerable threats to the public and at a time when conventional regulatory programs clearly have substantial technical and administrative problems.

NOTES

1. *Technologies and Management Strategies for Hazardous Waste Control*, Office of Technology Assessment, Washington, D.C., March 1983.

2. "Review of Activities of Major Firms in the Commercial Hazardous Waste Management Industry: 1981 Update," Booz-Allen & Hamilton, U.S. EPA, U.S. Government Printing Office, Washington, D.C., May 7, 1982.

11
Toward a Stable Treatment Market: The Role of Industrial Lobbying

BRUCE PIASECKI

It has been five years since Love Canal was declared a national emergency, yet the United States still lacks an effective chemical treatment market. In contrast with the space age detoxification practices of other nations, U.S. policy continues to favor dumping toxic wastes in landfills.

The rationale most frequently given by the Environmental Protection Agency and the generators of toxic wastes is that alternatives to dumping are technologically premature and prohibitively expensive. Yet this response is utterly at variance with the example set by European countries and with recent improvements in the U.S. detoxification arsenal.

We could employ these safe new detoxification devices and thereby retrieve valuable resources, decrease fuel costs, and close loops in production systems. But investors and corporate decision makers are not sure, due to the continued policy difficulties studied in the first part of this book, if they have a reliable market.

Asked if treatment is faltering from the recession, Allan Farkas, a consultant for EPA, says, "The decreasing number of acceptable land disposal sites for increasing volumes of toxic wastes makes competition in the treatment market bullish. Yet few American companies can respond to this strong demand with state-of-the-art solutions."

What, then, has kept toxic waste treatment the nation's biggest infant

Parts of this chapter first appeared in an article entitled "Struggling to Be Born," in the spring 1983 issue of *Amicus Journal*.

Although much of the original personnel of HWTC has changed since the writing of this chapter, its philosophy and purposes have continued to reflect a need for better and stronger controls in hazardous waste management. For an update on the growth and current priorities of the council, please see Appendix C, written by the Executive Director of HWTC, Richard C. Fortuna.

industry? Since Love Canal, environmentalists have argued that the answer is nothing less than a lack of strong and consistent regulations. The story behind the nation's inability to shift from dumping to treatment is a classic tale of how regulatory neglect can lead to market failure.

Ironically, the most forceful evidence for stronger toxic waste controls has come most recently from the corporate sector. Companies that had tried to capture the treatment market stimulated by the Carter EPA, and that began getting the cold shoulder from the Reagan EPA, are now fighting back. According to attorney Marvin B. Durning, a former representative of the Hazardous Waste Treatment Council (HWTC), "Early in 1982, a month before Anne Gorsuch [now Anne Burford] lifted the ban against dumping liquid toxic wastes, a significant number of American companies came to our office in Washington, D.C. to oppose the Reagan EPA. They feared federal backsliding of toxic waste controls. They were right." The HWTC incorporated that same month and began filing suit against EPA.

This alliance with environmental groups holds much hope for the future, for it gives economic clout to concerns about public health. HWTC's representation already includes ENSCO, SCA Services, and CECOS—three of the top seven American firms handling 70 percent of the toxic waste treatment market.

HWTC AND THE LIQUIDS CONTROVERSY

The HWTC's first major battle was to preserve a long-standing ban against dumping toxic liquids into landfills, which had been lifted by Burford in February 1982. HWTC lawyers met with congressional leaders and the press, explaining why EPA's suspension of the ban was ill conceived. (See chapter 3 for a more thorough account of these issues.)

Understanding the fierce logic of this problem, HWTC lawyers argued that 11 of the 19 states managing major commercial landfills (including New York, Colorado, Michigan, and Louisiana) had implemented bans against dumping toxic liquids by the time Burford reversed the trend. Industrial documents disclosed that thousands of companies employed safe dewatering techniques. SCA Services, one of the big seven waste management firms, decided in 1981 against dumping select wastes altogether after evaluating the hazards of liquid leachates. The HWTC issued a judicial appeal for EPA to preserve the ban. The results of these efforts, together with public outcry and heavy congressional pressure, led to the reimposition of the ban in late March.

However, two weeks after Burford opened landfills to liquids, IT Corporation announced that it was no longer in the business of building high-tech integrated detoxification systems. IT had been a leader in waste treatment for some time, with an impressive staff of experts at its Envirosciences offices in Knoxville, Tennessee. But Burford's move carried the implicit message that waste treatment services were not needed. HWTC formed to fight the obvious question her actions had given rise to: Why should IT and its competitors invest in treatment technologies when they can continue cheap dumping?

Asked if IT's decision to renege on high-tech detoxification was unique, Allan Farkas said "No, not at all. Others will follow suit. Although IT did not yet have the rotary-kiln incinerators ENSCO and Rollins used, nor the sophisticated chemical facilities of SCA Services, they were a leading treatment company with a strong potential for growth."

Faced with the prospect of losing other affiliates like IT, the HWTC increased its efforts to keep EPA from aborting the treatment market once again. It launched a massive lobbying effort to close long-standing loopholes in the Resource Conservation and Recovery Act (RCRA), the major federal legislation meant to prevent toxic contamination by encouraging treatment and recovery. The result was bountiful industrial evidence that EPA should strengthen its bite.

THE BOILER EXEMPTION

Testifying before the House Subcommittee on Commerce, Transportation, and Tourism after the reimposition of the ban, Durning issued a new warning. Every year about 20 million metric tons (about one third to one half of the annual volume of waste) are burned in facilities, such as industrial boilers, completely unregulated under RCRA.

Since EPA relies on RCRA rather than the Clean Air Act to protect the public from toxic air emissions from incineration, this long-standing loophole known as the "boiler exemption" became the focus of the HWTC's second major campaign.

Issued to relieve the energy crisis in 1978, the loophole totally exempts toxic wastes "being burned as a fuel for the purpose of recovering usable energy" from treatment and disposal standards. Under carefully controlled conditions, wastes can be used as fuel provided their incineration is 99.99 percent efficient. But trade group representatives knew the boiler exemption was being used to dispose of toxic wastes. Under the boiler exemption, one New Jersey oil blender began selling home heating oil laced with high levels of PCBs to unsuspecting consumers in Manhattan. The practice was first documented by the Chemical Manufacturers Association (CMA) in September 1981 when CMA officials filed evidence with EPA that 43 percent of boilers it had surveyed did not come close to meeting safety standards for incineration. EPA sat on this evidence for months. In April of 1982, HWTC notified Burford it intended to sue EPA for permitting this hazard to stand. By June, New York and Michigan had filed similar lawsuits.

Arguing that EPA "has made factual findings that incineration of wastes must achieve 99.99 percent destruction and removal efficiency if it is to be safe," HWTC lawyers used EPA's own yardstick to explain why the boiler exemption presents an unacceptable level of risk. Citing internal documents from a well-attended EPA workshop on boilers, HWTC lawyers listed examples of widespread abuse, from pier 91 in Seattle to residential areas in Westbury, Connecticut.

In recent months, HWTC has lobbied for passage in the House of a bill to close the boiler exemption. The HWTC also has argued for requiring treatment of certain wastes prior to disposal. In each case, it has joined industrial experience with skillful legal and technical demonstration on how regulatory reform secures safe treatment options.

HWTC's greatest contribution may be that it has shattered a long-standing logjam between the industries involved with toxic wastes and the policymakers regulating the industry. If these two forces speak openly and share information effectively, environmentalists can expect a more workable climate for charting the circuitous path toward the safer management of toxic wastes.

OTHER INDUSTRIAL EFFORTS

Clearing the air of the suspicions that surround the toxic waste crisis will take years; yet already evidence is mounting that the logjam is loosening elsewhere. Once again, the subject of industry's wrath has been a loophole in RCRA that preempts the budding treatment market. This time it is the landfillers themselves who are fighting for stronger regulations.

Issued by EPA for reasons of "administrative convenience," the small generators exemption allows 695,000 or 92 percent of toxic waste generators to slip through the net of federal controls. Under this exemption generators can dump at any solid waste landfill and municipal dump without records or monitoring, as long as they dump less than one ton of toxic wastes per month. A recent report issued by the Office of Technology Assessment indicates that more than 5 percent of toxic wastes are exempt from all federal controls under this clause.

Realizing that 1,000 kilograms per month is a large volume of deregulated poisons, 20 states have undertaken separate action to restrict or close this loophole. Although EPA had stated in its initial May 1980 exemption that it would require a reduction from 1,000 kilograms to 100 kilograms per month within two to five years, the Reagan EPA reneged and announced before Senate subcommittees that it would not restrict small generators at all. The decision flew in the face of the waste disposal industry. Solid waste handlers were suffering horrible accidents resulting from toxic wastes disposed of with regular trash; hundreds of incidents of unexpected explosions in shipping and worker burns were known to have resulted from small generators. Sixty-five of the first 115 contaminated sites identified by EPA for emergency cleanup were solid waste facilities and municipal dumps, most of which had accepted untreated toxic wastes from small generators.

As a result, the National Solid Waste Management Association (NSWMA), the primary trade group representing the firms that run landfills, opposed EPA openly and petitioned Congress to force the public's environmental watchdog to strengthen its bite. Eleven state chapters of NSWMA joined forces to explain to the public how much more consistent and better enforced toxic controls would help the economy and preserve public health.

Speaking before the Association of Petroleum Re-refiners, Representative Florio summed up this new industrial challenge to the Reagan EPA by saying: "Unless the administration issues and enforces hazardous waste regulations in a meaningful way, the same free market forces that brought us Love Canal will continue to proliferate. The technologies are here and now and have demonstrated ability, but they will not grow or displace unsound practices unless there are controls that ensure everyone is playing by the same rules."

Afterword

Unfouling the Nest: New Detoxification Strategies

BRUCE PIASECKI

In the late 1970s, residents of Hardeman County, Tennessee, began to suffer nausea and vomiting, to have trouble breathing, and eventually, to have children with birth defects; their drinking water had been contaminated by pesticides from a nearby dump. Several years ago at a farm in Hemlock, Michigan, not far from a toxic waste disposal area, chicks were hatched with organs outside their bodies. In Elizabeth, New Jersey, discarded drums of benzene, carbon tetrachloride, and other cancer-causing chemicals exploded on April 21, 1980, rattling windows across the Hudson River in downtown Manhattan.

This list of disasters keeps growing,[1] and will continue to grow, until we begin restricting the land disposal of select toxic wastes. This is the primary message made often enough in this book, but there is another side to this story. Major treatment methods capable of defusing America's worst wastes remain stalled in industrial laboratories for lack of market incentives and regulatory encouragement. What do these detoxification devices look like, and how might they begin to rectify America's problems with toxic waste? Here are three examples.

MOLTEN SALT INCINERATION

It has been known for a few years that a good soak in a hot bath of molten salt will destroy DDT powder and certain agents of chemical

This chapter was developed from an article, "Unfouling the Nest," which appeared in the September issue of *Science 83*. Reprinted by permission of *Science 83* Magazine, copyright the American Association for the Advancement of Science.

warfare such as mustard gas. Recent studies done for the EPA[2] concluded that the process also can destroy toxic solvents and acids.

For this process, waste is injected into a pool containing sodium carbonate heated to about 1,650 degrees Fahrenheit. (The hot salt, often laced with corrosive wastes, tended to devour the walls in early models; in later units the walls were made more resistant.) Hazardous hydrocarbons immediately burn, turning into carbon dioxide and water vapor, while substances like sulfur and chlorine react with the salt to form an easily managed inorganic ash. Molten salt's great efficiency is achieved at temperatures below that of conventional incineration. The lower temperature means that nitrous oxides, which are created in conventional high-temperature burning and contribute to acid rain, are minimized.

Less than one percent of the waste's original volume is left after a trip through the pool of salt, while toxic substances are destroyed well beyond the standard of excellence that EPA researchers call the "four nines"—99.99 percent.

Rockwell International offers a unit that can detoxify one ton of waste per hour as well as smaller, transportable models that destroy 10 to 250 pounds per hour. "But why should a company employ our device," says project manager Jim Johanson, "when they can burn toxic wastes without meeting safety standards or dump wastes directly into imperfect landfills?"[3] The fate of a dozen innovations in thermal destruction[4] hangs upon the closure of long-standing loopholes, such as the boiler exemption described in chapter 11, that allow the inexpert burning of toxic wastes.

PCB TRUCKS

By 1970 it was clear that polychlorinated biphenyls, or PCBs, spelled trouble. Research tied PCBs to birth defects, skin lesions, and tumors—and inventories were turning up evidence of PCBs everywhere. As an efficient medium of heat transfer, PCBs were mixed with oil and used as cooling fluid in millions of utility transformers and other electrical equipment. In 1970, U.S. production of PCBs peaked at 85 million pounds.

In 1977, Congress banned further production,[5] but that still left the country with more than a billion gallons of contaminated oil inside a variety of electrical components. Burying the contaminated materials safely was too expensive. Cleansing the oil of its PCBs was considered the better strategy. That same year, numerous research and development directors faced the same question: Since PCBs are unlikely to degrade in the soil, how might we best design a device that can destroy these substances known for their persistent chemical stability?

Goodyear Tire and Rubber Company unveiled the first process in 1979. It wasn't a moment too early; around the same time, the EPA reported that 91 percent of a cross section of Americans tested already had detectable levels of PCBs in their fatty tissue.[6] But Goodyear left the commercialization of the process to others. Within a few months of each other in 1980, two companies devised a similar approach, and one of them, Sun Ohio Company, housed their process in a 40-foot truck so that it could be moved from place to place.

These innovations occurred in anticipation of regulations restricting the dumping of PCBs. In July of 1981, an EPA report stated, "An estimated 20 million pounds of PCBs are now being stored awaiting proper disposal; about another 10 million pounds continue to leak, spill, and evaporate into the environment yearly."[7]

Soon after, Sun Ohio informed the EPA that it was ready to employ its device. After decontaminating transformer fluids containing PCBs in excess of 10,000 parts per million, the Sun Ohio trucks fulfilled EPA's tests for quality control in the summer of 1981.[8] Once again, major innovations had occurred without a moment to spare. On March 1, 1981, the EPA had banned the land disposal of PCB-contaminated liquids and electric utility equipment containing over 500 parts per million. The evidence of leaking landfills was simply too strong to permit further dumping of PCBs in America.

Without an ultimate sink to dump PCBs, technicians now take these complex, man-made substances apart with incisive precision. The contaminated oil is pumped into a chemical solution that contains metallic sodium. The sodium strips the chlorine from the PCBs and combines with it, creating a form of sodium chloride similar to common table salt. The biphenyls are removed as a harmless sticky residue, and the oil can be returned to the transformer, used as fuel oil, or dumped without special handling. The process can reduce PCB contamination from a level of 10,000 parts per million to less than 2 parts per million.[9]

Impressed by Sun Ohio's system of filtering, dehydrating, and degassing by-products, the EPA once again strengthened its bite.[10] The 20 million pounds of PCBs now in storage must be detoxified. To meet this challenge, numerous companies are building their own fleet of PCB-destroyer trucks.

Acurex's 1983 model, for instance, deletes the use of naphthalene, a hazardous substance found in mothballs and often employed in original sodium reagent mixtures. Equipped with a gas chromatograph to ensure consistent peak destruction, the Acurex process also destroys PCBs to less than two parts per million—obliterating PCBs to less than one twenty-fifth of concentrations health officials consider acceptable.

Presently, Acurex is developing a related process involving dechlorination that may help rid sites such as Times Beach, Missouri, of dioxin that has poisoned its soils. One truckload at a time, the contaminated soil would be dug up, fed into the portable treatment plant, and then returned to the ground.[11]

The lesson in destroyer-truck innovations[12] is obvious but usually overlooked. Once we eliminate the dead end of trying to figure out the best way to dump a certain poison, engineers devise new ways to turn them into harmless chemicals. The pace of this innovation quickens whenever regulations ensure someone a profit from investing in these high-tech solutions.

PLASMA ARC DESTRUCTION

Plasma arc detoxification obliterates wastes beyond our most sensitive gauge of parts per trillion. These plasma torches, adapted for toxic waste destruction by Canadian engineer Tom Barton of the Royal Military College of Canada in Kingston, Ontario, provide the highest destructive energy possible, short of atomic reactors.[13]

Barton's invention sends a powerful bolt of electricity between two electrodes, heating the air in the two-foot-long chamber to a temperature of 45,000 degrees Fahrenheit and creating an intense stampede of energy known as the plasma state.[14]

Sometimes called a fourth state of matter, plasma is a cloud of high-energy atoms, ions, and electrons. Because many of the particles in a plasma are charged, it is an extremely efficient conductor of electricity. As much as 99 percent of all matter in the universe—including stars, the aurora borealis, and interstellar space—is in a plasma state. Lightning bolts, fluorescent lights, and welding arcs are also plasma-like. In Barton's machine, toxic waste is fed into the center of the chamber where high-energy electrons bombard the material, breaking molecular bonds and reducing the substance to its basic elements, typically carbon, hydrogen, oxygen, and chlorine.

NASA developed plasma arcs 20 years ago to test the ability of space vehicles to face the intense heat of reentry. These powerful arcs consumed their electrodes in minutes. What's new about Barton's approach is that the electrodes can withstand a sustained plasma state, thus his device is reuseable and capable of a continuous feed of toxic substances.

Plasma devices can detoxify waste chemicals six million times faster than burning. Moreover, the destructive range of plasma arcs appears almost limitless. Barton has fed PCB-contaminated substances, in concentrations of 22,000 parts per million, into his unit and come out with no traceable evidence of PCB. On other runs, Barton has destroyed a wide spectrum of common pesticides and the ubiquitous carcinogen carbon tetrachloride.

In October of 1982, 60 industries interested in plasma technology met for a symposium.[15] They concluded that plasma arcs can destroy solid, liquid, or gas wastes, both organic and inorganic. Experts at Rockwell and Battelle Memorial Labs confirm that plasma arcs have the potential for detoxifying this vast domain. Westinghouse may use similar devices to melt waste steel, and the U.S. Army is looking hard at plasma science to destroy obsolete nerve gases.

After 23 years of a NASA career associated with developing plasma tunnels to test manned spacecraft, John E. Grimaud, head of the L.B.J. Space Center's Heat Transfer Section, reviewed Barton's process for the U.S. EPA. Writing in September of 1982 to Herbert L. Wiser, the EPA's director of environmental engineering, Grimaud condoned Barton's system as "technically sound . . . with a very high probability of performing."[16] With the go-ahead from L.B.J., federal EPA reviews are proceeding promptly at the time this book goes to press.

Yet it was the low cost of plasma destruction that brought the state of New York to Barton's door.[17] Faced with the burden of spending a million dollars to build a second collection facility to store the thousands of gallons of toxic leachate still leaking from the Love Canal site, the state signed a $754,000 contract with Barton in July of 1983.[18] His first target will be the 2,000 gallons of leachate already collected in storage tanks since the site was declared a national emergency. Barton will provide the state with compact trucks that detoxify 60 gallons of waste per hour. This marks a major step beyond dumping in America.

CONCLUSION

There are no tricks in toxic waste management. Advances such as Barton's, or those by the dozens of designers behind molten salt and PCB-dechlorination trucks, occur slowly and seldom with magical ease. Yet the ever-widening menace of leaking landfills can now begin to be stopped.

In the race to prevent future Love Canals, the devices named here are only three among the front contenders. For every molten salt unit, major innovations in thermal destruction wait. For every PCB truck, a team of relatives is ready to roll. For every Tom Barton, other world-class runners seek their chance to qualify. (See appendix A for a list of other available technologies.)

Right now, A. M. Chakrabarty, the former GE microbiologist who designed the first oil-eating microbes, is genetically engineering superbugs that can decompose highly obstinate wastes. In his lab at the University of Illinois in Chicago, he inserts plasmids—little circles of DNA—into bacteria. On these plasmids, he has spliced genes capable of producing enzymes that degrade dioxin—one of the most poisonous substances made by man.

Michael Modell, an MIT engineer turned entrepreneur, has designed a detoxification device aptly called "supercritical water," which pushes water to a point of intense high pressure and temperature. Suddenly it becomes a devastating solvent, which dissolves toxic wastes flowing daily from pesticide, plastic, and electronic industries.[19]

The remarkable good news, often neglected in the wake of leaking landfills, is that detoxification is possible. It is a thriving, economical alternative to dumping but one that requires a battery of policymakers and managers charged with a severe distrust of inexpert disposal. For without leadership committed to moving America beyond dumping, even these life-saving machines are only half of a pair of scissors.

NOTES

1. For a more thorough account of these leaking landfills, see Michael Brown's *Laying Waste: The Poisoning of America by Toxic Chemicals* (reviewed, with citations, in annotated bibliography). Also, see the following articles by Michael Brown: "Poison in Hemlock," *Quest/80*, September 1980, pp. 34–38, 102; "The Price of Life," *New York Magazine*, March 10, 1980, pp. 28–32; and "Killer Towns," *Rolling Stone*, November 26, 1981, pp. 31–32, 34, 78.
2. For a detailed technical assessment of molten salt destruction, see Harry M. Freeman, "Molten Salt Destruction Process: An Evaluation for Application in California," EPA/California Cooperative Agreement Papers—EPA R808908, August 1981. (45 pages.) For a copy, write: E. Timothy Oppelt, Industrial Environmental Research Laboratory, U.S. EPA, Cincinnati, Ohio 45268. Also see OTA report, *Technologies and Management Strategies for Hazardous Waste Control*, pp. 170–72 (citation in annotated bibliography).
3. Interview with Jim Johanson, June 18, 1983, Rockwell International.
4. For a technical overview of these promising new ways of burning toxic substances, see Harry M. Freeman, "Review and Evaluation of Selected Hazardous Waste Destruction Processes." (52 pages.) Obtained from same EPA source cited in note 2.
5. For a summary of PCB regulations, see "Polychlorinated Biphenyls (PCBs) Manufacturing, Processing, Distribution in Commerce and Use Prohibitions," *Federal Register*, May 31, 1979, vol. 44, no. 106.
6. "Controlling PCBs—A New Approach," *EPA Journal*, July/August 1981, vol. 7, no. 7, p. 25.
7. Ibid.
8. "Portable PCB Destruction Process Awaits EPA Approval," *Toxic Materials News*, September 1981, p. 300.
9. "PCBX Chemical Destruction of PCBs," as presented to the Annual Conference of the Southeastern Electric Exchange, Atlanta, Georgia, April 23, 1981. Contact: Norman E. Jackson, Sun Ohio, at 216–453–4677. Also, see *Chemical Week*, September 17, 1980, p. 56 and *Chemical Week*, April 29, 1981, p. 24.
10. U.S. EPA Region IX letter by Sonia F. Crow, regional administrator, to Norman E. Jackson, chairman of Sun Ohio, December 17, 1981. (3 pages.)
11. "Chemical Decomposition of PCBs: The Acurex Process," by G. James Mille, as presented at PCB Seminar at the Electric Power Research Institute, Dallas, Texas, December 1–3, 1981.
12. For more information on PCB destruction, see L. Weitzman and L. Pruce,

"Disposing Safely of PCBs: What's Available, What's on the Way," *Power*, February 1981, p. 78.

13. Contact: Dr. T. G. Barton, professor of engineering, Department of Civil Engineering, 2073 Sawyer Building, Royal Military College, Kingston, Ontario, Canada K7L 2W3, 613–545–7395.

14. T. G. Barton, "Composite Waste Atomization," presented at the Chemical Institute of Canada, 65th Canadian Chemical Conference, May 30–June 2, 1982, Toronto, Ontario. Also, see Barton, "Plasma Arc Pyrolysis of PCBs," presented at the International Hazardous Waste Conference, October 27–29, 1980, Denver, Colorado.

15. "Industrial Opportunities for Plasma Technology," proceedings of a symposium held by the Government of Ontario, October 21, 1982, Toronto. Copies available from: Ontario Hydro, Research Division, 800 Kipling Avenue, Toronto, Canada M8Z 5S4. Contact: Dr. G. R. Floyd or Dr. C. J. Simpson at 416–231–4111.

16. John E. Grimaud, letter to Herbert L. Wiser at the EPA, September 28, 1982. (2 pages, with technical evaluations attached.)

17. For a detailed economic evaluation of Barton's device, see "PCB Destruction Technology in North America," a report written by the M. M. Dillon consulting firm for the Environmental Protection Service of Canada. Pages 89–91 review Barton's plasma arcs. Contact: T. A. Constantine, project manager for M. M. Dillon at 416–482–5656.

18. "Plasma Contract R–1, Contract Documents and Specifications: Evaluation of Innovative Technology for the Destruction of Hazardous Wastes," submitted to Henry Williams, Commissioner, on January 25, 1983. (384 pages, with supporting documents.) Contact: Henry Williams, New York State Department of Environmental Conservation, 50 Wolf Road, Albany, New York 12233.

19. For a technical assessment of supercritical water, see "Executive Summary: Ten-Year Technology Plan," by CECOS International, submitted to State of New York DEC Office, pp. 26–33. Contact: Frank R. Nero, president, CECOS International, Inc., Niagara Falls Blvd. and Walmore Road, Niagara Falls, New York, 716–731–3281.

Appendix A
Some Alternative Technologies: Excerpts from the Office of Technology Assessment Report

Hazardous Waste Control Assessment Project Staff

Lionel S. Johns, *Assistant Director, OTA*
Energy, Materials, and International Security Division

Audrey Buyrn, *Materials Program Manager*

Project Staff

Joel S. Hirschhorn, *Project Director*
Suellen Pirages, *Senior Analyst*
Karen Larsen, *Senior Analyst*
Iris Goodman, *Research Analyst*
William F. Davidson
Linda Curran (OTA Fellow)
Richard Parkinson
Frances Sladek
Mimi Turnbull

Administrative Staff

Carol A. Drohan   Patricia A. Canavan

Individual Contractors

Christopher T. Hill   David M. Kagan   James Farned   David Burmaster   David Conn
Suzanne Contos   Kitty Gillman   Gary Davis   J. Stedman Stevens

Contractors

Engineering Science
Association of State and Territorial Solid Waste Management Officials
Citizens for a Better Environment
Sterling-Hobe
American Technology Management Consultants
Environmental Law Institute
Bracken & Baram
Clement Associates
Hazardous Material Control Research Institute
Stanford Knowledge Integration Laboratory
Development Sciences, Inc.
Aspen Systems Corp.
Adams and Wojick Associates

I thank Joel Hirschhorn, director of OTA's report, for permission to reprint the following excerpts on waste reduction, recycling, and detoxification.—Ed.

For an overview of the entire OTA study, see the review under *Technologies and Management Strategies for Hazardous Waste Control* in the Annotated Bibliography.

Hazardous Waste Control Assessment Advisory Panel

Sam Gusman, *Chairman,* Senior Associate, Conservation Foundation

David G. Boltz
Director, Solid Waste Control
Environmental Control Division
Bethlehem Steel Corp.

Frank Collins
Technology and Public Affairs Consultant

Stacy L. Daniels
Environmental Sciences Research Laboratory
DOW Chemical, U.S.A.

Jeffrey Diver
Senior Environmental Counsel
Waste Management, Inc.

Phillipa Foot
Department of Philosophy
University of California, Los Angeles

Morton R. Friedman
Director, Research
Hazardous Waste and Resource Recovery
N.J. Department of Environmental Protection

Thomas H. Goodgame
Director of Corporate Environmental Control
Research and Engineering Center
Whirlpool Corp.

Diane T. Graves
Conservation Chairman
Sierra Club, New Jersey Chapter

Rolf Hartung
Professor of Toxicology
School of Public Health
University of Michigan

Robert Judd
Director, California Office of Appropriate Technology

Kenneth S. Kamlet
Director, Pollution and Toxic Substances Division
National Wildlife Federation

Terry R. Lash
Independent Consultant

David Lennett
Attorney
Environmental Defense Fund

Janet Matey
Chemical Manufacturers Association

Randy M. Mott
Attorney
Hazardous Waste Treatment Council

John M. Mulvey
Director of Engineering Management Systems
School of Engineering/Applied Science
Princeton University

Delbert Rector
Chief of Environmental Services
Michigan Department of Natural Resources

Gerard Addison Rohlich
LBJ School of Public Affairs
University of Texas at Austin

Reva Rubenstein
Manager of the Institute of Chemical Waste Management
National Solid Wastes Management Association

Bernard L. Simonsen
Vice President
IT Corp.

George M. Woodwell
Director of The Ecosystems Center
Marine Biological Laboratory

NOTE: The Advisory Panel provided advice and comment throughout the assessment, but the members do not necessarily approve, disapprove, or endorse the report for which OTA assumes full responsibility.

Technologies for Hazardous Waste Management

Summary Findings

Waste Reduction Alternatives

- **Source segregation** is the easiest and most economical method of reducing the volume of hazardous waste. This method of hazardous waste reduction has been implemented in many cases, particularly by large industrial firms. Many opportunities still exist for further application. Any change in management practices should include the encouragement of source segregation.
- Through a desire to reduce manufacturing costs by using more efficient methods, industry has implemented various **process modifications.** Although a manufacturing process often may be used in several plants, each facility has slightly different operating conditions and designs. Thus, a modification resulting in hazardous waste reduction may not be applicable industrywide. Also,proprietary concerns inhibit information transfer.
- **Product substitutes** generally have been developed to improve performance. Hazardous waste reduction has been a side-benefit, not a primary objective. In the long term, end-product substitution could reduce or eliminate some hazardous wastes. Because many different groups are affected by these substitutions, there are limitations to implementation.
- With regard to **recovery and recycling** approaches to waste reduction, if extensive recovery is not required prior to recycling a waste constituent, in-plant operations are relatively easy. Commercial recovery benefits are few for medium-sized generators. No investment is required, but liability remains with the generator. Commercial recovery has certain problems as a profitmaking enterprise. The operator is dependent on suppliers' waste as raw material; contamination and consistency in composition of a waste are difficult to control. Waste exchanges are not very popular at present, since generators must assume all liability in transferring waste. Also, small firms do not generate enough waste to make it attractive for recycling.

Hazard Reduction Alternatives: Treatment and Disposal

- Many waste treatment technologies can provide **permanent, immediate,** and very **high degrees** of hazard reduction. In contrast, the long-term effectiveness of land-based disposal technologies relies on **continued** maintenance and integrity of engineered structures and proper operation. For wastes which are toxic, mobile, persistent, and bioaccumulative, and which are amenable to treatment, hazard reduction by treatment is generally preferable to land disposal. In general, however, costs for land disposal are comparable to, or lower than, unit costs for thermal or chemical treatment.
- For waste disposal, advanced landfill designs, surface impoundments, and injection wells are likely to perform better than their earlier counterparts. However, there is insufficient experience with these more advanced designs to predict their performance. Site- and waste-specific factors and continued maintenance of final covers and well plugs will be important. The ability to evaluate the effectiveness of these disposal technologies could be improved through better instrumentation of these facilities. Currently, their performance evaluation relies heavily on monitoring the indirect effects of their failure by, for example, detecting aquifer contamination.
- In comparing waste treatment to disposal alternatives, the degrees of permanent hazard reduction immediately achievable with treatment technologies are overwhelming attributes in comparison to land-based dispos-

al. However, comparison of these technologies at the very high destruction levels they achieve is difficult. Difficulties include: monitoring methods and detection limits, knowledge about the formation of toxic products of incomplete combustion, and diversity in performance capabilities among the different treatments.

- Chemical, physical, and biological **batch-type treatment processes** can be used to reduce waste generation or to recover valuable waste-stream constituents. In marked contrast to both incineration and land disposal, these processes allow checking treatment residuals before any discharge to the environment. In general, processes which offer this important added reliability are few, but waste-specific processes are emerging. **Research and development efforts could encourage the timely emergence of more of this type process applicable to future hazardous wastes.**

Ocean Use

- For some acids and very dilute other hazardous wastes, dumping in ocean locations may offer acceptable levels of risk for both the ocean environment and human health. However, there is generally inadequate scientific information for decisions concerning most toxic hazardous wastes and most locations. This is a serious problem since there may be increasing interest in using the oceans as the costs of land disposal increase and if public opposition to siting new treatment facilities continues.

Uncontrolled Sites

- A major problem is that the National Contingency Plan under the Comprehensive Environmental, Response, Compensation, and Liability Act (CERCLA) does not provide specific standards, such as concentration limits for certain toxic substances, to establish the extent of cleanup. There are concerns that cleanups may not provide protection of health and environment over the long term.
- The long-term effectiveness of remedial technologies is uncertain. A history of effectiveness has not yet been accumulated. Many remedial technologies consist of waste containment approaches which require long-term operation and maintenance. In recent remedial actions, removal of wastes and contaminants, such as soil, accounted for 40 percent of the cases; such removed materials were usually land disposed.

Introduction

The purpose of this chapter is to describe the variety of technical options for hazardous waste management. The technical detail is limited to that needed for examining policy options and regulatory needs. Still, there are many technologies, and their potential roles in hazardous waste management are diverse. Thus, there are many technical aspects related to policy and regulation issues. The reader interested in the details of the technologies reviewed here is encouraged to read beyond this policy-oriented discussion.

The first group of technologies discussed are those which reduce waste volume. This distinction recognizes that where technically and economically feasible, it is better to reduce the generation of waste than to incur the costs and risks of managing hazardous waste. **Waste reduction** technologies include segregation of waste components, process modifications, end-product substitutions, recycling or recovery operations, and various emerging technologies. Many waste reduction technologies are closely linked to manufacturing and involve proprietary information. Therefore, there is less detailed information in this section than in others.

Much of the chapter discusses technologies that reduce the hazard from the waste generated. These are grouped as: 1) those **treatments that permanently eliminate the hazardous character** of the material, and 2) those **disposal approaches that contain** or immobilize the hazardous constituents.

There are several treatments involving high temperature that decompose materials into harmless constituents. Incineration is the obvious example, but there are several existing and emerging "destruction" technologies that are distinguished in this category. In addition to gross decomposition of the waste material, there are emerging chemical technologies which detoxify by limited molecular rearrangement and recover valuable materials for reuse. Whether by destruction or detoxification, these technologies permanently eliminate the hazard of the material.

Containment chiefly involves land disposal techniques, but chemical "pretreatment" methods for stabilization on a molecular level are rapidly emerging. Combining these methods offers added reliability, and sectors of industry appear to be adopting that approach. The discussion of containment technologies includes: 1) landfilling, 2) surface impoundments, 3) deep-well injection, and 4) chemical stabilization.

Use of the oceans is considered a technical option for some wastes. A number of regulatory and policy issues emerge concerning ocean use and are discussed.

The final section of this chapter concerns uncontrolled hazardous waste sites from which releases of hazardous materials is probable or has already occurred. Such sites are often abandoned and are no more than open dumps. The sites are addressed by CERCLA. The technical aspects of identifying, assessing, and remediating uncontrolled sites are reviewed in this section. There has been limited engineering experience with cleaning up uncontrolled sites.

Many technologies that are applicable to the same waste compete in the marketplace. The initial discussion in the section on hazard reduction treatment and disposal technologies compares the costs of comparative technologies in some detail.

Waste Reduction Alternatives

Introduction

Four methods are available to reduce the amount of waste that is generated:

1. source segregation or separation,
2. process modification,
3. end-product substitution, and
4. material recovery and recycling.

Often, more than one of these approaches is used, simultaneously or sequentially.

Reduction of the amount of waste generated at the source is not a new concept. Several industrial firms have established in-house incentive programs to accomplish this. One example is the 3P program—Pollution Prevention Pays—of the 3M Corp. Through the reduction of waste and development of new substitute products for hazardous materials, 3M has saved $20 million over 4 years.[1] Other firms have established corporate task forces to investigate solutions to their hazardous waste management problems. One solution has been recycling and recovery of waste generated by one plant for use as a raw material at another corporate-owned facility. Such an approach not only reduces waste, but lowers operating costs.

Significant reductions in the volume of waste generated can be accomplished through source segregation, process modification, end-product substitution, or recovery and recycle. No one method or individual technology can be selected as the ultimate solution to volume reduction. As shown in figure 7, three of the methods, i.e., source segregation, process modifica-

[1] M. G. Royston, *Pollution Prevention Pays* (New York: Pergamon Press, 1979).

Figure 7.—Relative Time Required for Implementation of Reduction Methods

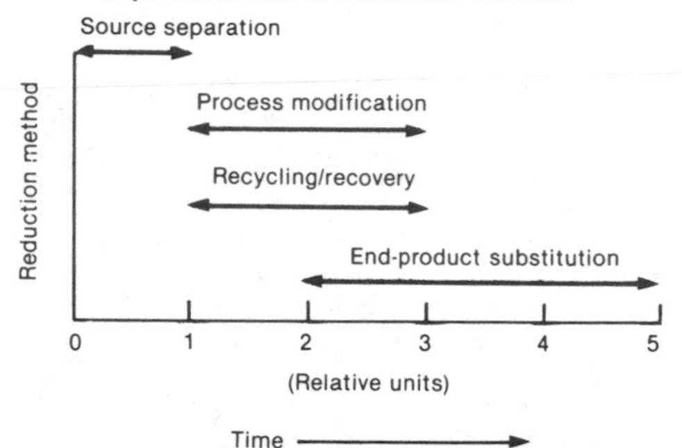

SOURCE: Office of Technology Assessment.

tions, and recovery and recycling can be implemented on a relatively short- to medium-term basis by individual generators. End-product substitution is a longer term effort. A comparison of the advantages and disadvantages of each of the four approaches is given in table 22. Because of proprietary concerns and lack of industrywide data, the amount of waste reduction that has already occurred and the potential for further reduction is difficult to evaluate. A 1981 study by California[2] concluded that new industrial plants will produce only half the amount of hazardous waste currently produced. Other estimates for potential waste production range from 30 to 80 percent.[3] Waste reduction efforts, however, are more difficult in existing plants.

Source Segregation

Source segregation is the simplest and probably the least costly method of reduction. This approach prevents contamination of large vol-

[2]"Future Hazardous Waste Generation in California," Department of Health Services, Oct. 1, 1982.

[3]Joanna D. Underwood, Executive Director, Inform, *The New York Times*, Dec. 27, 1982.

Table 22.—A Comparison of the Four Reduction Methods

Advantages	Disadvantages
Source segregation or separation	
1) Easy to implement; usually low investment	1) Still have some waste to manage
2) Short-term solution	
Process modification	
1) Potentially reduce both hazard and volume	1) Requires R&D effort; capital investment
2) Moderate-term solution	2) Usually does not have industrywide impact
3) Potential savings in production costs	
End product substitution	
1) Potentially industrywide impact—large volume, hazard reduction	1) Relatively long-term solutions
	2) Many sectors affected
	3) Usually a side benefit of product improvement
	4) May require change in consumer habits
	5) Major investments required—need growing market
Recovery/recycling	
• *In-plant*	
1) Moderate-term solution	1) May require capital investment
2) Potential savings in manufacturing costs	2) May not have wide impact
3) Reduced liability compared to commercial recovery or waste exchange	
• *Commercial recovery (offsite)*	
1) No capital investment required for generator	1) Liability not transferred to operator
	2) If privately owned, must make profit and return investment
2) Economy of scale for small waste generators	3) Requires permitting
	4) Some history of poor management
	5) Must establish long-term sources of waste and markets
	6) Requires uniformity in composition
• *Waste exchange*	
1) Transportation costs only	1) Liability not transferred
	2) Requires uniformity in composition of waste
	3) Requires long-term relationships—two-party involvement

SOURCE: Office of Technology Assessment.

umes of nonhazardous waste by removal of hazardous constituents to form a concentrated hazardous waste. For example, metal-finishing rinse water is rendered nonhazardous by separation of toxic metals. The water then can be disposed through municipal/industrial sewage systems.

However, there are disincentives, particularly for small firms wishing to implement source segregation. For example, an electroplating firm may, for economic reasons, mix wastes containing cyanides and toxic metals with a waste that contains organics. The waste stream is sent to the municipal treatment system. The municipal system can degrade the organics, but the metals and cyanide accumulate in the sludge, which is disposed as a nonhazardous solid waste in a sanitary landfill. As long as the firm dilutes the cyanide and metals concentrations to acceptable limits for municipal disposal, it is in compliance with the Environmental Protection Agency's (EPA) regulations. If the firm calculates the costs of recovering the cyanide onsite, the cost may be more than the fees paid to the municipal treatment facility. Thus, there is no economic incentive for source segregation, which would yield a hazardous waste, although the public would benefit if source segregation were practiced. Alternatively, accumulation of such sludges can lead to significant levels of toxic material in sanitary landfills. Municipal treatment facilities are financed with tax dollars. In this example, the public is, in essence, subsidizing industrial waste disposal. Moreover, to carry out source segregation, a firm may have to invest in new equipment.

Process Modification

Process modifications are, in general, made on a continuous basis in existing plants to increase production efficiencies, to make product improvements, and to reduce manufacturing costs. These modifications may be relatively small changes in operational methods, such as a change in temperature, in pressure, or in raw material composition, or may involve major changes such as use of new processes or new equipment. Although process modifications have reduced hazardous wastes, the reduction usually was not the primary goal of the modifications. However, as hazardous waste management costs increase, waste reduction will become a more important primary goal.

Three case examples were studied to analyze incentives and impacts for process modifications for hazardous waste reduction. The following factors are important:

- A typical process includes several steps. Although a change in one step may be small relative to the entire process, the combination of several changes often represents significant reductions in cost, water use, or volume of waste.
- A change in any step can be made independently and is evaluated to determine the impact on product, process efficiency, costs, labor, and raw materials.
- Generally, process modifications are plant- or process-specific, and they cannot be applied industrywide.
- A successful process change requires a detailed knowledge of the process as well as a knowledge of alternative materials and processing techniques. Successful implementation requires the cooperative efforts of material and equipment suppliers and in-house engineering staffs.

Three process changes are discussed in detail in the appendix to this chapter and are briefly summarized below:

1. **Chlor-alkali industry.**—Significant process developments in the chlor-alkali industry (which produces, e.g., chlorine and caustic soda) have resulted in reduction of major types of hazardous waste through modifications to the mercury electrolysis cell. The effects on waste generation are summarized in table 23. The modifications were not developed exclusively to reduce hazardous waste, but were initiated primarily to increase process efficiency and reduce production costs.
2. **Vinyl chloride (plastics) production.**—Several process options are available for handling waste from the production of vinyl chloride monomers (VCMs). Five al-

Table 23.—Process Modifications to the Mercury Cell

Modification	Effect on waste stream	Reason for modification
Diaphragm cell	Elimination of mercury contaminated waters	Preferred use of natural salt brines as raw material
Dimensionally stable anode	Elimination of chlorinated hydrocarbon waste	Increased efficiency
Membrane cell	Elimination of asbestos diaphragm waste	Reduce energy costs; higher quality product

SOURCE: Office of Technology Assessment.

ternatives are illustrated in table 24. All five have been demonstrated on a commercial scale. In most cases, the incineration options (either recycling or add-on treatment) would be selected over chlorinolysis and catalytic fluidized bed reactors. Chlorinolysis has limited application because of the lack of available markets for the end products. If further refinements could be made to the catalytic process, such as higher concentration of hydrogen chloride (HCl) in the gas stream which would allow it to be used with all oxychlorination plants, its use could be expanded.

3. **Metal-finishing industry.**—Several modifications in metal cleaning and plating processes have enabled the metal-finishing industry to eliminate requirements for onsite owned and operated wastewater treatment facilities. By changing these processes to eliminate formation of hazardous sludge, the effluent can be discharged directly to a municipal wastewater treatment facility, saving several million dollars in capital investment.

End-Product Substitution

End-product substitution is the replacement of hazardous waste-intensive products (i.e., industrial products the manufacture of which involves significant hazardous waste) by a new product, the manufacture of which would eliminate or reduce the generation of hazardous waste. Such waste may arise from the ultimate disposal of the product (e.g., asbestos products) or during the manufacturing process (e.g., cadmium plating).

Table 25 illustrates six examples of end-product substitution, each representing a different type of problem. General problems include the following:

Table 24.—Advantages and Disadvantages of Process Options for Reduction of Waste Streams for VCM Manufacture

Treatment option	Type	Advantages	Disadvantages
High-efficiency incineration of vent gas only	Add-on treatment	1. Relatively simple operation 2. Relatively low capital investment	1. Second process required to handle liquid waste stream
High-efficiency incineration without HCl recovery	Add-on treatment	1. Relatively simple operation 2. Relatively low capital investment	1. Loss of HCl
High-efficiency incineration with HCl recovery	Recycling	1. Heat recovery 2. Recover both gaseous and liquid components 3. High reliability	1. Exit gas requires scrubbing 2. Requires thorough operator training 3. Auxiliary fuel requirements
Chlorinolysis	Modification of process	1. Carbon tetrachloride generated	1. High temperatures and pressures required 2. High capital investment costs 3. Weakening market for carbon tetrachloride
Catalytic fluidized bed reactor	Recycling	1. Low temperature 2. Direct recycle of exit gas (no treatment required)	1. Limited to oxychlorination plants

SOURCE: Office of Technology Assessment.

Table 25.—End-Product Substitutes for Reduction of Hazardous Waste

Product	Use	Ratio of waste:[a] original product	Available substitute	Ratio of waste:[a] substitute product
Asbestos	Pipe	1.09	Iron Clay PVC	0.1 phenols, cyanides, 0.05 fluorides 0.04 VCM manufacture + 1.0 PVC pipe
	Friction products (brake linings)	1.0+ manufacturing waste	Glass fiber Steel wool Mineral wools Carbon fiber Sintered metals Cement	0
	Insulation	1.0+ manufacturing	Glass fiber Cellulose fiber	0.2
PCBs	Electrical transformers	1.0	Oil-filled transformers Open-air-cooled transformers	0 0
Cadmium	Electroplating	0.29	Zinc electroplating	0.06
Creosote treated wood	Piling		Concrete, steel	0.0 (reduced hazard)
Chlorofluorocarbons	Industrial solvents	70/81 = 0.9	Methyl chloroform; methylene chloride	0.9 (reduced hazard)
DDT	Pesticide	1.0+ manufacturing waste	Other chemical pesticides	(reduced hazard) 1.0+ manufacturing waste

[a]Quantity of hazardous waste generated/unit of product

SOURCE: Office of Technology Assessment.

- **Not all of the available substitutes avoid the production of hazardous waste.** For example, in replacing asbestos pipe, the use of iron as a substitute in pipe manufacturing generates waste with phenols and cyanides; and also, during the manufacture of polyvinyl chloride (PVC) pipe, a hazardous vinyl chloride monomer is emitted.[4]
- **Substitution may not be possible in all situations.** For example, although a substantial reduction in quantity of hazardous waste generated is achieved by using clay pipes, clay is not always a satisfactory replacement for asbestos.[5]

Generally, development of substitutes is motivated by some advantage, either to a user, (e.g., in improved reliability, lower cost, or easier operation), or to the manufacturer (e.g., reduced production costs). A change in consumer behavior also may cause product changes. For example, increased use of microwave ovens has increased the demand for paper and Styrofoam packaging to replace aluminum. Most end-product substitutions aimed at reduced generation of hazardous waste, however, do not have such advantages. The **only** benefit may be reduction of potential adverse effects on human health or the environment. Unless the greater risks and costs of hazardous waste management are fully internalized by waste generators, other incentives may be needed to accomplish end-product substitution.

In addition to the approach in chapter 3, option III, end-product substitution may be encouraged by:

1. regulations,
2. limitation of raw materials,
3. tax incentives,
4. Federal procurement practices, and
5. consumer education.

Regulations have been used to prohibit specific compounds. For example, bans on certain pesticides such as dichlorodiphenyltrichloro-

[4]Sterling-Hobe Corp., *Alternatives for Reducing Hazardous Waste Generation Using End-Product Substitution,* prepared for Materials Program, OTA, 1982.

[5]Ibid.

ethane (DDT) have resulted in development and use of other chemicals. Legislative prohibition of specific chemicals, such as polychlorinated biphenyls (PCBs), is another option.

Limiting the supply of raw materials required for manufacture is another method of encouraging end-product substitution. For example, limiting either the importation or domestic mining of asbestos might encourage substitution of asbestos products. A model for this method is the marketing-order system of the U.S. Department of Agriculture, used to permit the cultivation of only specified quantities of selected crops. Using similar strategies, a raw material like asbestos could be controlled by selling shares of a specified quantity of the market permitted to be mined or imported.

Tax incentives are another means to force end-product substitution. Excise taxes on products operate as disincentives to consume and have been implemented in the past (e.g., taxes on alcohol, cigarettes and gasoline). This type of taxation might be incorporated to encourage product substitution. The design and acceptance of a workable, easily monitored tax system, however, might be difficult to develop.

Federal procurement practices and product specifications can have significant influence on industrial markets. Changes in military procurement were proposed in 1975 to allow for substitution of cadmium-plating by other materials. A change in product specifications to permit this substitution would affect not only the quantity of cadmium required for military use, but also might impact nonmilitary applications.

A public more aware of the hazard associated with production of specific products might be inclined to shift buying habits away from them.

Larger Economic Contexts.—If a substitution requires a complete shift in industrial markets (e.g., if a product manufactured with asbestos is replaced by one made with cement), the impact may be large—both manufacturers and suppliers may be affected. In addition, users will be impacted according to the relative merits of the products. Other sectors potentially affected by end-product substitution include importers of raw materials, exporters of the original product, and related equipment manufacturers.

Generally, a product substitution offers a cost advantage over the original product, which counters market development expenditures. Potential savings can be achieved by the introduction of product substitutes—e.g., increased demand may require increased production, thus reducing the cost per unit. Incentives or the removal of disincentives, however, may be necessary to increase product demand by a sufficient margin to give the substitute a more competitive marketplace position.

A significant factor in the introduction of a substitute product is the stage of growth for existing markets. For example, if the market for asbestos brake lining is declining or growing at a very slow rate, or if large capital investments are required for development of a substitute lining, introduction of a substitute may not be economically practical. The availability of raw materials also affects the desirability of substitutes. If the original product is dependent on limited supplies of raw materials, substitutes will be accepted more rapidly.

Recovery and Recycling

Recovery of hazardous materials from process effluent followed by recycling provides an excellent method of reducing the volume of hazardous waste. These are not new industrial practices. Recovery and recycling often are used together, but technically the terms are different. Recovery involves the separation of a substance from a mixture. Recycling is the **use** of such a material recovered from a process effluent. Several components may be recovered from a process effluent and can be recycled or discarded. For example, a waste composed of several organic materials might be processed by solvent distillation to recover halogenated organic solvents for recycling; the discarded residue of mixed organics might be burned for process heat.

Materials are amenable to recovery and recycling if they are easily separated from process effluent because of physical and/or chemical differences. For example, inorganic salts can be concentrated from aqueous streams by evaporation. Mixtures of organic liquids can be separated by distillation. Solids can be separated from aqueous solutions through filtration. Further examples of waste streams that are easily adaptable to recovery and recycling are listed in table 26.

Recovery and recycling operations can be divided into three categories:

1. **In-plant recycling** is performed by the waste (or potential waste) generator, and is defined as recovery and recycling of raw materials, process streams, or byproducts for the purpose of prevention or elimination of hazardous waste. (Energy recovery without materials recovery is not included in this discussion of in-plant recycling, but is discussed later in this chapter as a **treatment** of wastes.) If several products are produced at one plant by various processes, materials from the effluents of one process may become raw materials for another through in-plant recycling. An example is the recovery of relatively dilute sulfuric acid, which is then used to neutralize an alkaline waste. In-plant recycling offers several benefits to the manufacturer, including savings in raw materials, energy requirements, and disposal or treatment costs. In addition, by reducing or eliminating the amount of waste generated, the plant owner may be exempted from some or all RCRA (Resource Conservation and Recovery Act) regulations.
2. **Commercial (offsite) recovery** can be used for those wastes combined from several processes or produced in relatively small quantities by several manufacturers. Commercial recovery means that an agent other than the generator of the waste is handling collection and recovery. These recovery systems may be owned and operated by, or simply serve, several waste generators, thereby offering an advantage of economy of scale. In most cases commercial recovery systems are owned and operated by independent companies, and are particularly important for small waste generators. In commercial recovery, responsibility for the waste and compliance with regulations and manifest systems remains that of the generator until recovery and recycling is completed.
3. **Material exchanges** (often referred to as "waste" exchanges) are a means to allow raw materials users to identify waste generators producing a material that could be used. Waste exchanges are listing mechanisms only and do not include collection, handling, or processing. Although benefits occur by elimination of disposal and treatment costs for a waste as well as receipt of cash value for a waste, responsibility for meeting purchaser specifications remains with the generator.*

Standard technologies developed that can be adapted for recovery of raw materials or byproducts may be grouped in three general cate-

*For a discussion of the problems being encountered with using waste exchanges for hazardous waste see "Industrial Waste Exchange: A Mechanism for Saving Energy and Money," Argonne National Laboratory, July 1982.

Table 26.—Commercially Applied Recovery Technologies

Generic waste	Typical source of effluent	Recovery technologies
Solids in aqueous suspension	Salt/soda ash liming operations	Filtration
Heavy metals	Metal hydroxides from metal-plating waste; sludge from steel-pickling operations	Electrolysis
Organic liquids	Petrochemicals/mixed alcohol	Distillation
Inorganic aqueous solution	Concentration of inorganic salts/acids	Evaporation
Separate phase solids, grease/oil	Tannery waste/petroleum waste	Sedimentation/skimming
Chrome salt solutions	Chromium-plating solutions/tanning solutions	Reduction
Metals; phosphate sulfates	Steel-pickling operations	Precipitation

SOURCE: Office of Technology Assessment

gories. **Physical separation** includes gravity settling, filtration, flotation, flocculation, and centrifugation. These operations take advantage of differences in particle size and density. **Component separation** technologies distinguish constituents by differences in electrical charge, boiling point, or miscibility. Examples include ion exchange, reverse osmosis, electrolysis, adsorption, evaporation, distillation, and solvent extraction. **Chemical transformation** requires chemical reactions to remove specific chemical constituents. Examples include precipitation, electrodialysis, and oxidation-reduction reactions. These technologies are reviewed in table 27.

A typical recovery and recycling system usually uses several technologies in series. Therefore, what may appear as a complex process actually is a combination of simple operations. For example, recylcing steel-pickling liquors may involve precipitation, gravity settling, and flotation. Precipitation transforms a component of high solubility to an insoluble substance that is more easily separated by gravity settling, a coarse separation technique, and flotation, a finishing separation method. Integration of process equipment can introduce some complexity. The auxiliary handling equipment (e.g., piping, pumps, controls, and monitoring devices that are required to provide continuous treatment from one phase to another) can be extensive. A detailed description of the recycling and recovery of pickling liquors from the steel industry is provided in the appendix at the end of this chapter.

Recovery and recycling technologies applied to waste vary in their stages of development. Physical separation techniques are the most commonly used and least expensive. The separation efficiency of these techniques is not as high as more complex systems, and therefore the type of waste to which it is applied is limited. Complex component separations (e.g., reverse osmosis) are being investigated for application to hazardous waste. These generally are expensive operations and have not been implemented commercially for hazardous waste reduction. Chemical transformation methods are also expensive. Precipitation and thermal oxidation, however, appear to have current commercial application in hazardous waste management.

Table 28 illustrates some technologies currently being investigated for application to waste recovery and recycle. An expanded discussion of emerging new technologies, specifically in phase separation is provided in the following section of this chapter.

Economic Factors

These factors include:

1. research and development required prior to implementation of a technology;
2. capital investment required for new raw material, or additional equipment; i.e., recovery and recycle equipment, control equipment, and additional instrumentation;
3. energy requirements and the potential for energy recovery;
4. improvements in process efficiency;
5. market potential for recycled material, either in-house or commercially, and anticipated revenues;
6. management costs for hazardous waste before use of recovery and recycle technology;
7. waste management cost increases, resulting from recovery/recycling, i.e., additional manpower, insurance needs, and potential liability; and
8. the value of improved public relations of a firm.

Because of the number of processing steps involved, recovery and recycling can be more expensive than treatment and disposal methods. Earned revenue for recovered materials, however, may counter the cost of recovery.

Many market and economic uncertainties must be considered in an evaluation of proposed technology changes. For example, if deregulation of oil and natural gas results in an increase in energy costs, additional energy requirements, and/or credits earned for energy recovery from a process could be affected. The uncertainty of continued availability of a nec-

Table 27.—Description of Technologies Currently Used for Recovery of Materials

Technology/description	Stage of development	Economics	Types of waste streams	Separation efficiency[a]	Industrial applications
Physical separation:					
Gravity settling: Tanks, ponds provide hold-up time allowing solids to settle; grease skimmed to overflow to another vessel	Commonly used in wastewater treatment	Relatively inexpensive; dependent on particle size and settling rate	Slurries with separate phase solids, such as metal hydroxide	Limited to solids (large particles) that settle quickly (less than 2 hours)	Industrial wastewater treatment first step
Filtration: Collection devices such as screens, cloth, or other; liquid passes and solids are retained on porous media	Commonly used	Labor intensive: relatively inexpensive; energy required for pumping	Aqueous solutions with finely divided solids; gelatinous sludge	Good for relatively large particles	Tannery water
Flotation: Air bubbled through liquid to collect finely divided solids that rise to the surface with the bubbles	Commercial application	Relatively inexpensive	Aqueous solutions with finely divided solids	Good for finely divided solids	Refinery (oil/water mixtures); paper waste; mineral industry
Flocculation: Agent added to aggregate solids together which are easily settled	Commercial practice	Relatively inexpensive	Aqueous solutions with finely divided solids	Good for finely divided solids	Refinery; paper waste; mine industry
Centrifugation: Spinning of liquids and centrifugal force causes separation by different densities	Practiced commercially for small-scale systems	Competitive with filtration	Liquid/liquid or liquid/solid separation, i.e., oil/water; resins; pigments from lacquers	Fairly high (90%)	Paints
Component separation:					
Distillation: Successfully boiling off of materials at different temperatures (based on different boiling points)	Commercial practice	Energy intensive	Organic liquids	Very high separations achievable (99+% concentrations) of several components	Solvent separations; chemical and petroleum industry
Evaporation: Solvent recovery by boiling off the solvent	Commercial practice in many industries	Energy intensive	Organic/inorganic aqueous streams; slurries, sludges, i.e., caustic soda	Very high separations of single, evaporated component achievable	Rinse waters from metal-plating waste
Ion exchange: Waste stream passed through resin bed, ionic materials selectively removed by resins similar to resin adsorption. Ionic exchange materials must be regenerated	Not common for HW	Relatively high costs	Heavy metals aqueous solutions; cyanide removed	Fairly high	Metal-plating solutions
Ultrafiltration: Separation of molecules by size using membrane	Some commercial application	Relatively high	Heavy metal aqueous solutions	Fairly high	Metal-coating applications
Reverse osmosis: Separation of dissolved materials from liquid through a membrane	Not common: growing number of applications as secondary treatment process such as metal-plating pharmaceuticals	Relatively high	Heavy metals; organics; inorganic aqueous solutions	Good for concentrations less than 300 ppm	Not used industrially

Table 27.—Description of Technologies Currently Used for Recovery of Materials—Continued

Technology/description	Stage of development	Economics	Types of waste streams	Separation efficiency[a]	Industrial applications
Electrolysis:					
Separation of positively/negatively charged materials by application of electric current	Commercial technology; not applied to recovery of hazardous materials	Dependent on concentrations	Heavy metals; ions from aqueous solutions; copper recovery	Good	Metal plating
Carbon/resin absorption:					
Dissolved materials selectively absorbed in carbon or resins. Absorbents must be regenerated	Proven for thermal regeneration of carbon; less practical for recovery of adsorbate	Relatively costly thermal regeneration; energy intensive	Organics/inorganics from aqueous solutions with low concentrations, i.e., phenols	Good, overall effectiveness dependent on regeneration method	Phenolics
Solvent extraction:					
Solvent used to selectively dissolve solid or extract liquid from waste	Commonly used in industrial processing	Relatively high costs for solvent	Organic liquids, phenols, acids	Fairly high loss of solvent may contribute to hazardous waste problem	Recovery of dyes
Chemical transformation:					
Precipitation:					
Chemical reaction causes formation of solids which settle	Common	Relatively high costs	Lime slurries	Good	Metal-plating wastewater treatment
Electrodialysis:					
Separation based on differential rates of diffusion through membranes. Electrical current applied to enhance ionic movement	Commercial technology, not commercial for hazardous material recovery	Moderately expensive	Separation/concentration of ions from aqueous streams; application to chromium recovery	Fairly high	Separation of acids and metallic solutions
Chlorinolysis:					
Pyrolysis in atmosphere of excess chlorine	Commercially used in West Germany	Insufficient U.S. market for carbon tetrachloride	Chlorocarbon waste	Good	Carbon tetrachloride manufacturing
Reduction:					
Oxidative state of chemical changed through chemical reaction	Commercially applied to chromium; may need additional treatment	Inexpensive	Metals, mercury in dilute streams	Good	Chrome-plating solutions and tanning operations
Chemical dechlorination:					
Reagents selectively attack carbon-chlorine bonds	Common	Moderately expensive	PCB-contaminated oils	High	Transformer oils
Thermal oxidation:					
Thermal conversion of components	Extensively practiced	Relatively high	Chlorinated organic liquids; silver	Fairly high	Recovery of sulfur, HCl

[a]Good implies 50 to 80 percent efficiency, fairly high implies 80 percent, and very high implies 90 percent.

SOURCE: Office of Technology Assessment.

Table 28.—Recovery/Recycling Technologies Being Developed

Technology	Development needs	Potential application
Ion exchange	Commercial process for other applications (desalinization), applications to metal recovery under development. Not economic at present due to investment requirements	Chromium recovery; metal-plating waste
Adsorption	R&D on new resins and regeneration methods	Organic liquids with or without metal contamination; pesticides
Electrolysis	Cathode/anode, material development for membranes	Metallic/ionic solution
Extraction	Reduction in loss of acid or solvent in process	Extraction of metals with acids
Reverse osmosis	Membrane materials, operating conditions optimized, demonstration of process	Salt solutions
Evaporation	Efficiency improvement/ demonstration of process	Fluorides from aluminum smelting operation
Reduction	Efficient collection techniques	Mercury
Chemical dehalogenation	Equipment development for applications to halogenated waste other than PCB oils	Halogenated organics

SOURCE: Office of Technology Assessment.

essary raw material could influence a decision for recovery of materials from waste streams. Uncertainties in interest rates may discourage investment and could thus increase a required rate-of-return projected for a new project. Changes in allowable rates of capital equipment depreciation also may affect costs significantly.

In addition, changes in RCRA regulations for alternative management options (e.g., landfilling, ocean dumping, and deep-well injection) affect disposal costs. Stricter regulations or prohibitions of certain disposal practices for particular wastes could increase the attractiveness of recycling and recovery operations. However, if hazardous wastes are stored for longer than 90 days, current regulations require permits for that facility. If large quantities of a waste must accumulate (for economic reasons) prior to recycling or recovery, the permit requirement may discourage onsite recycling.

Previously, recovery and recycling was considered as an in-plant operation only; i.e., material was recovered and recycled within one plant. Currently, larger corporations are beginning to evaluate recovery opportunities on a broader scale. Recycling within the corporate framework is gaining greater attention as a cost reduction tool with an added benefit of reducing public health risks.

Emerging Technologies for Waste Reduction

Although the effects are more difficult to predict, some technological developments have potential for the reduction of hazardous waste. For example, developments in the electronics industry have provided instrumentation and control systems that have greater accuracy than was possible just a few years ago. These systems provide more precise control of process variables, which can result in higher efficiency and fewer system upsets, and a reduction in hazardous waste. The application and improvements of instrumentation and control systems vary with each process. Thus, as new plants are constructed and fitted with new technologies, smaller quantities of hazardous waste will be generated. The technologies that are discussed in this section have a direct impact on the volume and hazard level of waste currently generated through one or more of the reduction methods discussed earlier.

Segregation Technology.—New developments in segregation technology can increase recovery and recycling of hazardous waste. Notably, membrane segregation techniques have substantially improved. Membrane separation has been used to achieve filtration, concentration, and purification. However, large-scale applications, such as those required in pollution control have been inhibited by two factors: 1) replacement costs associated with membrane use and 2) technical difficulties inherent in producing large uniform surface areas of uniform quality. Because of the inherent advantages of membrane separation over more conventional separation techniques like distillation or evaporation, further development of membrane separation for large-scale commercial applications is attractive. These advantages include lower energy requirements resulting in reduced operating costs and a simpler, more compact system that generally leads to reduced capital costs. Commercial applications exist for all but coupled transport designs, which are still at the laboratory stage. All of these illustrated systems have possible application for reduction of hazardous waste. However, microfiltration, ultrafiltration, reverse osmosis, and electrodialysis processes have more immediate application. Dialysis has been used on only a small scale; the high flow systems generally typical of hazardous waste treatments make its use impractical. Gas separations by membranes do not have immediate application to hazardous waste use. The development of new materials for both membranes and supporting fabrics and the use of new layering techniques (e.g., composite membranes) have led to improved permeability and selectivity, higher fluxes, better stability, and a reduced need for prefiltering and staged separations.

Improved reliability is the most important factor in advancement of membrane separations technology. New types of membranes have demonstrated improved performance. Thin-film composites that can be used in reverse osmosis, coupled transport, and electrolytic membranes have direct application to the recovery and reduction of hazardous materials from a processing stream.

The major cost in a membrane separation system is the engineering and development work required to apply the system to a particular process. Equipment costs are secondary; membranes generally account for only 10 percent of system costs. However, membranes must be replaced periodically and sales of replacement membranes are important to membrane production firms. Currently the largest profit items are for high-volume flow situations (e.g., water purification) or for high-value product applications (e.g., pharmaceutical productions). Over 20 companies cover the membrane market; the largest company is Millipore with 1980 total sales of $255 million.

The predicted market growth rate for membrane segregations is healthy, generally 10 to 20 percent annually of the present membrane market ($600 million to $950 million). Chloralkali membrane electrodialysis cells for the production of chlorine and sodium hydroxide lead the projected application areas in hazardous waste with growth rates of 25 to 40 percent of the present market ($10 million to $15 million). The recovery of chromic acid from electroplating solutions by coupled transport also has direct application for the reduction of hazardous waste. Other uses include ultrafiltration of electrocoat-painting process waste and waste water recovery by reverse osmosis. The use of membrane segregation systems in pretreatment of hazardous waste probably is the largest application for the near future.

Biotechnology.—Conventional biological treatments have been used in industrial waste treatment systems for many years (see tables 29 and 30). Recent advances in the understanding of biological processes have led to the development of new biological tools, increasing the opportunities for biotechnology applications in many areas, including the treatment of dilute hazardous waste. The potential impacts of these advancements on waste treatment techniques, process modifications, and end-product substitutes are discussed here.

Biotechnology has direct application to waste treatment systems to degrade and/or detoxify

Table 29.—Conventional Biological Treatment Methods

Treatment method	Aerobic (A) anaerobic (N)	Waste applications	Limitations
Activated sludge	A	Aliphatics, aromatics, petrochemicals, steelmaking, pulp and paper industries	Volatilization of toxics; sludge disposal and stabilization required
Aerated lagoons	A	Soluble organics, pulp and paper, petrochemicals	Low efficiency due to anaerobic zones; seasonal variations; requires sludge disposal
Trickling filters	A	Suspended solids, soluble organics	Sludge disposal required
Biocontactors	A	Soluble organics	Used as secondary treatment
Packed bed reactors	A	Nitrification and soluble organics	Used as secondary treatment
Stabilization ponds	A&N	Concentrated organic waste	Inefficient; long retention times, not applicable to aromatics; sludge removal and disposal required
Anaerobic digestion	N	Nonaromatic hydrocarbons; high-solids; methane generation	Long retention times required; inefficient on aromatics
Landfarming/spreading	A	Petrochemicals, refinery waste, sludge	Leaching and runoff occur; seasonal fluctuations; requires long retention times
Composting	A	Sludges	Volatilization of gases, leaching, runoff occur; long retention time; disposal of residuals

Aerobic—requires presence of oxygen for cell growth
Anaerobic—requires absence of oxygen for cell growth
SOURCE: Office of Technology Assessment.

Table 30.—Industries With Experience in Applying Biotechnology to Waste Management

Industry	Effluent stream	Major contaminants
Steel	Coke-oven gas scrubbing operation	NH_3, sulfides, cyanides, phenols
Petroleum refining	Primary distillation process	Sludges containing hydrocarbons
Organic chemical manufacture	Intermediate organic chemicals and byproducts	Phenols, halogenated hydrocarbons, polymers, tars, cyanide, sulfated hydrocarbons, ammonium compounds
Pharmaceutical manufacture	Recovery and purification solvent streams	Alcohols, ketones, benzene, xylene, toluene, organic residues
Pulp and paper	Washing operations	Phenols, organic sulfur compounds, oils, lignins, cellulose
Textile	Wash waters, deep discharges	Dyes, surfactants, solvents

SOURCE: Office of Technology Assessment.

chemicals. Development of new microbial strains can be used to improve:

1. degradation of recalcitrant compounds,
2. tolerance of severe or frequently changing operating conditions,
3. multicompound destruction,
4. rates of degradation, and
5. ability to concentrate nondegradable constituents.

Compounds thought to be recalcitrant, (e.g., toluene, benzene, and halogenated compounds) have been shown to be biodegradable by isolated strains. Strain improvement in these species through genetic manipulations has lead to improved degradation rates. Opportunities exist for applications of this technology in remedial situations—i.e., cleanup at spills or abandoned sites.[6] [7] The improvement of conventional biological systems through the development of specific microbial strains ("superbugs") capable of degrading multiple compounds has been proposed. However, this approach faces engineering difficulties, and development of collections of organisms working together might be preferable.

Development of **biological pretreatment systems** for waste streams has some potential for those wastes that contain one or two recalcitrant compounds. A pretreatment system designed to remove a specific toxic compound could reduce the shock effects on a conventional treatment process. In some cases, a pretreatment system may be used with other nonbiological treatment methods (i.e., incineration) to remove toxic compounds that may not be handled in the primary treatment system or to make them more readily treated by the primary system. In other cases, pretreatment might render a waste nonhazardous altogether.

One area of research in advanced plant genetics is in the use of plants to accumulate metals and toxic compounds from contaminated soils. Current research is direct to four areas. The first involves use of plants to decrease the metal content of contaminated soils, through increased rates of metal uptake. Plants then could be used to decontaminate soils through concentration of compounds in the plant fiber. The plants then would be harvested and disposed. The second area of development focuses on direct metal uptake in nonedible portions of the plant. For example, the development of a grain crop like wheat that could accumulate metal from soil in the nonusable parts of the plant would allow commercial use of contaminated land. A third area of research is directed toward development of crops that can tolerate the presence of metal without incorporating these toxic elements in plant tissue. Finally, research is being conducted concerning the use of plants in a manner similar to microorganisms to degrade high concentrations of hazardous constituents.

Changes in process design incorporating advances in biological treatment systems may result in less hazardous waste. The development of organisms capable of degrading specific recalcitrant materials may encourage source separation, treatment, and recycling of process streams that are now mixed with other waste streams and disposed. The replacement of chemical synthesis processes with biological processes may result in the reduction of hazardous waste. Two methods of increasing the rate of chemical reactions are through higher temperatures and catalysts. One type of catalyst is biological products (enzymes) that inherently require milder, less toxic conditions than do other catalytic materials.

Historically, many biological processes (fermentations) have been replaced by chemical synthesis. Genetic engineering offers opportunities to improve biological process through reduced side reactions, higher product concentrations, and more direct routes; thus, genetic engineering offers a means of partially reversing this trend. The development of new process approaches would require new reactor designs to take advantage of higher biological reaction rates and concentrations.

Biotechnology also could lead to substitution of a less or nonhazardous material for a hazardous material, particularly in the agricultural field. One of the primary thrusts of plant genetics is the development of disease-resistant plants, thus reducing the need for commercial products such as fungicides. Genetic engineering to introduce nitrogen-fixation capabilities within plants could reduce the use of chemical fertilizers and potentially reduce hazardous waste generated in the manufacture of those chemicals. However, two problems must be resolved before large-scale applications: 1) the genetic engineering involved in nitrogen fixa-

[6] G. T. Thibault and N. W. Elliott, "Biological Detoxification of Hazardous Organic Chemical Spills," in *Control of Hazardous Material Spills*, Conference Proceedings (Nashville, Tenn.: Vanderbilt University, 1980), pp. 398-402.

[7] G. C. Walton and D. Dobbs, "Biodegradation of Hazardous Materials in Spill Situations," in *Control of Hazardous Material Spills*, Conference Proceedings (Nashville, Tenn.: Vanderbilt University, 1980), pp. 23-45.

tion is complex and not readily achieved, and 2) the overall energy balance of internal nitrogen-fixation may reduce growth rates and crop yield.

Major Concerns for Biotechnology.—Although genetic engineering has some promising applications in the treatment of hazardous waste streams, several issues need to be addressed prior to widespread commercialization of the technology:[8]

- The factors for scale-up from laboratory tests to industrial applications have not been completely developed. Limited field tests have shown degradation rates in the field may be much slower than laboratory rates where pure cultures are tested in pure compounds.
- Basic biochemical degradation mechanisms are not well understood. The potential exists for the formation of other hazardous compounds through small environmental changes or system upsets and, without this basic understanding, chemical pathways cannot be anticipated.
- The potential exists for release of hazardous compounds into the environment through incomplete degradation or system failure.
- There is a possibility of adverse effects resulting from the release of "engineered" organisms into the environment.

The potential benefits of applied genetics to hazardous waste probably outweigh these factors. Although these factors must be addressed, they should motivate rather than overshadow research in this area.

Chemical Dechlorination With Resource Recovery.—In the late 1970's private efforts were undertaken to find a reagent that would selectively attack the carbon-chlorine bond under mild conditions, and thus chemically strip chlorine from PCB-type chemicals forming a salt and an inert sludge. Goodyear Tire & Rubber Co. made public its method. Sunohio and Acurex Inc. have developed proprietary reagents, modified the process, and commercialized their processes with mobile units. These processes reduce the concentration of PCB in transformer oil, which may be 50 to 5,000 parts per million (ppm) to less than 2 ppm. The Sunohio PCBX process is used for direct recycling of the transformer oil back into transformers, while the oil from the Acurex process is used as a clean fuel in boilers.[9]

Although, the development of these processes was initially aimed at PCB-laden oils of moderate concentration (50 to 500 ppm), their chemistry is generic in that it attacks the carbon-halogen bonds under mild conditions. Thus, they are potentially applicable to pesticides and other halogenated organic wastes as well as wastes with higher concentrations of PCBs. The PCBX process has been applied to pesticides and other halogenated waste with detoxification observed, but without published numerical results or further developments.[10] Acurex claims it has commercially treated oil with a PCB concentration of 7,000 ppm. In tests performed by Battelle Columbus Laboratories for Acurex, its process reduced dioxin concentration in transformer oil from 380 parts per trillion (ppt) to 40 ± 20 ppt. Acurex and the Energy Power Research Institute have tested the effectiveness of the process in the laboratory on capacitors which contain 100 percent PCB (40 to 50 percent chlorine, by weight). The next step is construction of a mobile commercial-scale facility which would shred, batch process, and test the capacitor material.[11]

The Sunohio (first to have a chemical dechlorination process approved by EPA) has five units in operation. Acurex has four mobile units in operation at this time and at least two other companies currently market similar chemical PCB destruction services. Acurex,

[8] S. P. Pirages, L. M. Curran, and J. S. Hirschhorn, "Biotechnology in Hazardous Waste Management: Major Issues," paper presented at *The Impact of Applied Genetics on Pollution Control*, symposium sponsored by the University of Notre Dame and Hooker Chemical Co., South Bend, Ind., May 24-26, 1982.

[9] *Alternatives to the Land Disposal of Hazardous Wastes*, Governor's Office of Appropriate Technology, California, 1981.

[10] Oscar Norman, developer of the PCBX process, personal communication, January 1983.

[11] Leo Weitzman, Acurex Corp., personal communication, January 1983.

Sunohio, and licensees have been selling their PCB services for over a year. Acurex and The Franklin Institute plan to commercialize their processes for spill sites involving halogenated organics.[12]

[12] Charles Rogers, Office of Research and Development, Industrial and Environmental Research Laboratory (IERL), Environmental Protection Agency, Cincinnati, Ohio, personal, communication, January 1983.

As an alternative to incineration, these chemical processes offer the advantages of no air emissions, no products of incomplete combustion, reduced transportation risks, and the recycling of a valuable material or the recovery of its fuel value. Further, as with many chemical processes, there is the opportunity to directly check the degree of destruction before any product is discharged or used.

Hazard Reduction Alternatives: Treatment and Disposal

Introduction

The previous section discussed technologies to reduce the volume of waste generated. This section analyzes technologies that reduce the hazard of waste. These include **treatment** and **disposal technologies. These two groupings of technologies contrast distinctly in that it is preferable to permanently reduce risks to human health and the environment by waste treatments that destroy or permanently reduce the hazardous character of the material, than to rely on long-term containment in land-based disposal structures.**

In the United States, as much as 80 percent (by volume) of the hazardous waste generated is land disposed (see ch. 4). Of these wastes, a significant portion could be treated rather than land disposed for greater hazard reduction. In California, for example, wastes which are toxic, mobile, persistent and bioaccumulative comprise about 29 percent of the hazardous waste disposed of offsite.[13] [14]

Following a brief summary comparison, this section reviews over 15 treatment technologies. Many of these eliminate the hazardous character of the waste. Technologies in the next group discussed are disposal alternatives. Their effectiveness relies on containing the waste to prevent, or to minimize, releases of waste and human and environmental exposure to waste. In this category, the major techniques are landfills, surface impoundments, and underground injection wells.

[13] California Department of Health Services, "Initial Statement of Reasons for Proposed Regulations (R-32-82)," Aug. 18, 1982, p. 23.

[14] California Department of Health Services, "Current Hazardous Waste Generation," Aug. 31, 1982, p. 6.

This discussion begins with a comparison of the treatment and disposal technologies and ends with a cost comparison. These discussions focus on the competitive aspects of the numerous hazard reduction technologies. However, choosing among these technical alternatives involves consideration of many factors, some of which are neither strictly technical or economic. Choices by waste generators and facility operators also depend on Federal and State regulatory programs already in place, those planned for the future, and on perceptions by firms and individuals of existing regulatory burdens may exist for a specific waste, technology, and location.

Summary Comparison

For the purpose of an overview, qualitative comparisons among technologies can be made. Based on principle considerations relevant across all technologies, the diverse range of hazard reduction technologies can be compared as presented in table 31. The table summarizes the important aspects of the above issues for each generic grouping of technologies included. Individual technologies are considered in more detail in the following discussions on treatment and disposal technologies. For simplicity, the technologies are grouped generically, and only a limited number

Table 31.—Comparison of Some Hazard Reduction Technologies

	Disposal		Treatment		
	Landfills and impoundments	Injection wells	Incineration and other thermal destruction	Emerging high-temperature decomposition[a]	Chemical stabilization
Effectiveness: How well it contains or destroys hazardous characteristics	Low for volatiles, questionable for liquids; based on lab and field tests	High, based on theory, but limited field data available	High, based on field tests, except little data on specific constituents	Very high, commercial-scale tests	High for many metals, based on lab tests
Reliability issues:	Siting, construction, and operation Uncertainities: long-term integrity of cells and cover, liner life less than life of toxic waste	Site history and geology; well depth, construction and operation	Long experience with design Monitoring uncertainties with respect to high degree of DRE; surrogate measures, PICs, incinerability	Limited experience Mobile units; onsite treatment avoids hauling risks Operational simplicity	Some inorganics still soluble Uncertain leachate test, surrogate for weathering
Environmental media most affected:	Surface and ground water	Surface and ground water	Air	Air	None likely
Least compatible waste:[b]	Liner reactive; highly toxic, mobile, persistent, and bioaccumulative	Reactive; corrosive; highly toxic, mobile, and persistent	Highly toxic and refractory organics, high heavy metals concentration	Possibly none	Organics
Costs: Low, **M**od, High	L-M	L	M-H (Coincin. = L)	M-H	M
Resource recovery: potential	None	None	Energy and some acids	Energy and some metals	Possible building material

[a]Molten salt, high-temperature fluid wall, and plasma arc treatments.
[b]Waste for which this method may be less effective for reducing exposure, relative to other technologies. Waste listed do not necessarily denote common usage.

SOURCE: Office of Technology Assessment.

of groups are compared. The principal considerations used for comparison are the following:

- **Effectiveness.**—This does not refer to the intended end result of human health and environmental protection, but to the capability of a technology to meet its specific technical objective. For example, the effectiveness of chemical dechlorination is determined by how completely chlorine is removed. In contrast, the effectiveness of landfills is determined by the extent to which containment or isolation is achieved.
- **Reliability.**—This is the consistency over time with which a technology's objective is met. Evaluation of reliability requires consideration of available data based on theory, laboratory-scale studies, and commercial experience.

A prominent factor affecting the relative reliability of a technology is the adequacy of **substitute performance measures.** Verification that a process is performing as designed is not always possible and, when possible, verification to a high level of confidence may require days or weeks to complete and may not be useful for timely adjustments. In some cases, key process variables can be used as substitute measures for the effectiveness of the technology. Substitute measures are used either because they provide faster and/or cheaper performance information. A disadvantage of surrogate measures is that there may not be reliable correlation between the surrogate measurement and the nature of any releases to the environment.

The reliability of a technology should also be judged on the **degree of process and discharge control available.** This refers to the ability to: 1) maintain proper operating conditions for the process, and 2) correct undesirable releases. Process control requires that information about performance be fed back to correct the process. Control systems vary categorically with respect to two important time variables:

1. the length of time required for information to be fed back into the system (e.g., time for surrogate sampling and analysis, plus time for corrective adjustments to have the desired effect); and
2. the length of time for release of damaging amounts of insufficiently treated materials in the event of a treatment upset.

In the case of landfills, once ground water monitoring has detected a leak, damaging discharges could have already occurred.

If detection systems are embedded in the liner, then detection of a system failure is quicker and more reliable, and it offers more opportunity for correction. Landfilling and incineration are examples where these time factors are important. In contrast, batch treatment processes, as discussed in the preceding section on "Waste Reduction," offer the distinct opportunity to contain and check any release, and retreat it if needed, so that actual releases of hazardous constituents are prevented. Other chemical and biological treatments are flow-through processes, with different rates of flow-through. These treatments vary in their opportunity for discharge correction. Generally, processes used in waste segregration and recycling offer this kind of reliability.

- **Environmental media most affected.**—This refers to the environmental media contaminated in the event that the technology fails.
- **Least compatible waste.**—Some technologies are more effective than others in preventing releases of hazardous constituents when applied to particular types of waste.
- **Costs.**—Costs vary more widely among generic groups of technologies than within these groups. Table 31 presents generalized relative costs among these groups. The final section of this chapter gives some unit management cost details.
- **Resource recovery potential.**—Treatments that detoxify and recover materials for recycling are discussed under "Waste Reduction." However, some materials, as well as energy, can be recovered with some of the technologies reviewed in this section. To the extent that materials and fuels are recovered and used, the generation of other hazardous wastes may be reduced. Potential releases of hazardous constituents from recovery and recycling operations must also be considered.

Treatment Technologies

In this section, treatment technologies refers to those techniques which decompose or break down the hazardous wastes into nonhazardous constituents.* Most of these treatments use high temperatures to decompose waste. Some of the promising emerging technologies cause decomposition by high-energy radiation and/or electron bombardment. There are several important attributes of high-temperature destruction technologies which make them attractive for hazardous waste management:

- the hazard reduction achieved is **permanent;**
- they are **broadly applicable to waste mixes;** most organics, for example, may be converted into nonhazardous combustion products; and
- the **volume** of waste that must ultimately be land disposed is greatly **reduced**.

In addition, with some of these treatments, there is a possibility of recovering energy and/or materials.** However, potential recovery of energy and materials is not the primary focus of this discussion.

Incineration is the predominant treatment technology used to decompose waste. The term "incineration" has been given a specific meaning in Federal regulations, where it denotes a particular subclass of thermal treatments, and draft Federal regulations may give specific meaning to the additional terms "industrial boiler" and "industrial furnace." Although the Federal definitions affect the manner in which a facility is regulated, unless specifically noted,

---

*Treatments can also be used to **segregate** specific waste constituents, or to **mitigate** their characteristics of ignitability, corrosiveness, or reactivity. Most of these are referred to as "industrial unit processes," and their use is usually embedded in larger treatment schemes. A lengthy listing will not be reproduced here. Many were described in the preceding section on "Waste Reduction Technologies." The interested reader is also referred to any industrial unit operations manual. Another source is "Chemical, Physical, Biological (CPB) Treatment of Hazardous Wastes," Edward J. Martin, Timothy Oppelt, and Benjamin Smith, Office of Solid Waste, U.S. Environmental Protection Agency, presented at the Fifth United States-Japan Governmental Conference of Solid Waste Management, Tokyo, Japan, Sept. 28, 1982.

**For example, the Chemical Manufacturers Association claims that a significant portion of the hydrochloric acid produced in the United States and some sulfuric acid come from incineration of chlorinated organics through wet-scrubbing of the stack gases. (CMA, personal communication, December 1982.) Also, there is clear potential for metals recovery with the emerging high-temperature technologies.

"combustion" or "incineration" are used in this report to refer to the generic processes of interest, and do not necessarily mean specific facility designs or regulatory categories.

Applicable Wastes

Liquid wastes are generally more easily incinerated than sludge or waste in granular form, because they can be injected easily into the combustion chamber in a manner which enhances mixing and turbulence. Wastes with heterogeneous physical characteristics and containerized or drummed wastes are difficult to feed into a combustion chamber. The rotary kiln is designed for sludge-like, granular and some containerized waste. Recently, a new firm has emerged (Continental Fibre Drum) which manufactures combustible fiber drums for waste containers. These fiber drums of organic waste can be incinerated in specially designed rotary kilns.

Elemental metals, of course, cannot be degraded. Waste which contain excessive levels of volatile metals may not be suitable for incineration. Under the high-temperature conditions in an incinerator, some metals are volatilized or carried out on particulates. Oxides of metals can generally be collected electrostatically. However, some volatilized forms cannot be electrically charged, resisting electrostatical collection. These include metallic mercury, arsenic, antimony, and cadmium, and very small particles.[15] (Particles having insufficient surface area also can't be adequately charged and collected.) Wet second-stage electrostatic precipitators are designed for removing these forms of volatized metals, but they are expensive and not in widespread use. High-pressure drop-emission controllers have also been effective, but their use is declining.

Technical Issues

There are approximately 350 liquid injection and rotary kiln incinerators currently in service for hazardous waste destruction.[16] Most of these facilities may eventually be permitted as RCRA hazardous waste incinerators. A far greater, although unknown, number of facilities may be combusting hazardous waste principally in order to recover their heating value. Under current regulations, these facilities would not be permitted as hazardous waste incinerators.[17] Under future regulations they may become subject to performance standards similar to those in effect for incinerators, be prohibited from burning certain types of ignitable hazardous waste, or be subject to some intermediate level of regulation.

To regulate incinerators, EPA has decided to use **performance standards** rather than specification of design standards. The current regulations specify three performance standards for hazardous waste incineration.[18] These standards are described below:

1. A 99.99 percent destruction and removal efficiency (DRE) standard for each principal organic hazardous constituent (POHC) designated in the waste feed. (This is the most difficult part of the standard to meet.) The DRE is calculated by the following mass balance formula:
   DRE = (1 − Wout/Win) × 100 percent,
   where:
   Win = the mass feed rate of 1 POHC in the waste stream going into the incinerator, and
   Wout = the mass-emission rate of the same POHC in the exhaust prior to release to the atmosphere.
2. Incinerators that emit more than 4 lb of hydrogen chloride per hour must achieve a removal efficiency of at least 99 percent. (All commercial scrubbers tested by EPA have met this performance requirement.)
3. Incinerators cannot emit more than 180 milligrams (mg) of particulate matter per dry standard cubic meter of stack gas. This standard is intended to control the emissions of metals carried out in the exhaust gas on particulate matter. (Recent tests indicate that this standard may be more difficult to achieve than was earlier thought.[19])

[15]Frank Whitmore, Versar, Inc., personal communication, August 1982.
[16]Gene Crumpler, Office of Solid Waste, Hazardous and Industrial Waste Division, Environmental Protection Agency, personal communication, January 1983.
[17]Ibid.
[18]40 CFR, sec. 264.343.
[19]Crumpler, op. cit.

There are instances in which the incinerator performance standards do not fully apply. First, the regulations do not apply to facilities that burn waste primarily for its fuel value. To date, energy recovery of the heat value of waste streams qualifies for the regulatory exemption.[20] Second, facilities burning waste that are considered hazardous because of characteristics of ignitability, corrosiveness, and reactivity are eligible for exemptions from the performance standards. Of the three, the exemption for energy recovery applies to a greater volume of hazardous waste. Finally, incinerators operating at sea are not governed by RCRA, but rather by the Marine Protection, Research, and Sanctuaries Act of 1972. Regulations under this act do not require scrubbing of the incinerator exhaust gas. In the future, EPA may require that incinerator ships operating in close proximity to each other scrub their exhaust gases.

With regard to combustion processes, the most important design characteristics are the "three Ts:"

1. maintenance of adequate **temperatures** within the chamber,
2. adequate **turbulence** (mixing) of waste feed and fuel with oxygen to assure even and complete combustion, and
3. adequate residence **times** in the high-temperature zones to allow volatilization of the waste materials and reaction to completion of these gases.

Finally, the DRE capability of these technologies generally varies widely depending on the waste type to which it is applied. Chlorine or other halogens in the waste tend to extinguish combustion; so, in general, these wastes tend to be more difficult to destroy. An important related misconception is that the more toxic compounds are the more difficult they are to burn. Toxic dioxins and PCBs are popular examples of highly halogenated wastes which are both highly toxic and difficult to destroy, but these should not imply a rule. Discussion of waste "incinerability" is included below.

[20]40 CFR, sec. 261.2 (c)(2).

Waste treatments with reliable high-destruction efficiencies offer attractive alternatives to land disposal for mobile, toxic, persistent, and bioaccumulative wastes. However, these treatment technologies are not free of technical issues. The first three issues noted below relate directly to **policy** and regulation, and the remaining three issues summarize sources of technical **uncertainty** with respect to the very small concentrations of remaining substances. Improvements in policy and regulatory control should recognize these technical issues:

- **Significant sources of toxic combustion products, emitted to the air, are not being controlled with the same rigor as are RCRA incinerators.** These include emissions from facilities inside the property boundaries of refineries and other chemical processing plant sites. In addition, "boilers" can receive and burn any ignitable hazardous waste which has beneficial fuel value (see discussion on "Boilers"). Draft regulations governing boilers are currently being developed under RCRA and very limited reporting requirements are brand new. Under the Clean Air Act, there is only very limited implementation governing the remaining facilities. Standards have been set for only four substances, and apply to only a small class of facilities.
- **There are some problems with the technology-based DRE performance standards.** EPA uses the technology-based performance standard for **practicality,** and for its **technology-forcing** potential. However, the performance standard overly simplifies the environmental comparisons among alternatives.

  Complete knowledge about the transport, fate, and toxic effects of each waste compound from each facility is unobtainable. Thus, some simplified regulatory tool is needed. However, the most important and known factors should be included in regulatory decisions. Notably, these could include: the **toxicity** of the waste, the load to the facility (the waste feed **concentra-**

tion and **size** of the facility), and **population** potentially affected. Future regulations, however, could endeavor to shape the manner in which competing technologies are chosen in a more environmentally meaningful way (see ch. 6).

Finally, the 99.99 percent DRE may be viewed as a "forcing" standard with respect to some high-temperature technologies, but emerging high-temperature technologies (notably plasma arc) may offer much greater and more reliable DREs. Rather than forcing, it may discourage the wide use of more capable technologies.

- **Strengthening regulations with respect to the technical uncertainties below will require deliberate research efforts in addition to anticipated permitting tests.** Test data for wastes that are difficult to burn are lacking. The current incinerator performance standard is based on EPA surveys from the mid-1970's which involved easier to burn wastes, higher fuel to waste feed ratios than in current use, and smaller than commercial-scale reactors. EPA is currently testing or observing test burns for many of the technologies described, using compounds found to be representative of very difficult to burn toxic waste. Most of these data are still being evaluated; few results have become available. In the next few years, a great deal of test burn data will be generated regarding existing facilities and given wastes. In addition, the cost of test burn is often $20,000 to $50,000. These costs burden both EPA research and private industry. Such data will help permit writers, but these data will have limited use in resolving many of the technical uncertainties described below.
- **Implementation of the current performance standards relies on industrywide use of monitoring technology operating at the limits of its capability.** In DRE analyses, the fourth nine is often referred to as guesswork; standardized stack gas sampling protocols for organic hazardous constituents are still being developed. This is particularly true with respect to organics carried on particulate matter and to the more volatile compounds. Methods for concentrating the exhaust gas in order to obtain the sample especially for volatile compounds are still evolving. The newness of these tests suggests there may be a wide variety in the precision capabilities among the laboratories which analyze DRE test results.
- **The measurements currently used in daily monitoring of performance cannot reliably represent DRE at the 99.99 percent level.** For recordkeeping and enforcement, air and waste feed rates along with gas temperature are used as indirect measures for DRE. For facilities already equipped with carbon monoxide meters (and for all Phase II regulated incinerators), carbon monoxide concentration in the stack gas is also included. Also, waste/fuel mix and waste/fuel ratio can have a great effect on DRE. Thus, these ratios are noted in the permits. However, it is difficult to specify acceptable ranges of mixes based on test burn information. The idea behind the specifications is that as long as actual values of these parameters remain within prescribed limits during operation, the desired DRE is being achieved. These measures are chosen not only because they are easily and routinely monitored, but also because there is a theoretical basis for using them to indicate combustion efficiency. However, all these measures are only indirectly related to the compounds of concern. For example, carbon monoxide is a very stable and easily monitored product of incomplete combustion (PIC). Thus, it is often used as a sensitive indicator for combustion efficiency in energy applications. However, its relationship to other combustion products and to remaining concentrations of POHCs is very indirect and uncertain.

Most experts agree that the development of a way to accurately measure DRE **concurrently** with treatment process, would eliminate much of the technical uncertainties surrounding incineration. To this end, EPA is studying devices which monitor total organic carbon, and the National

Bureau of Standards (NBS) is studying various combinations of available monitoring techniques.[21] It is not likely that a single technique can be developed in the near term to monitor the whole range of compounds of concern, but the development of a combination of devices to do the job holds promise. However, these techniques will still have problems. This will include: cost; some reliance on correlations to surrogate measures; and, in the case of the NBS approach, the possible introduction of corrosive tracer compounds.

- **There is sharp disagreement in the scientific and regulatory community about the use of waste "incinerability."** This concept is a regulatory creation, not a physical attribute of any material. The idea behind incinerability is that as long as the least incinerable waste (i.e., the most difficult to burn waste) is destroyed to the required extent, all other waste would be destroyed to an even greater extent. Thus, waste "incinerability," in addition to waste concentration, is used to select a limited number of waste constituents for monitoring in a **test burn.** Problems with this approach result largely from lack of basic information about measures for incinerability. This presents uncertainty in the selection of those POHCs to be monitored in the waste feed and stack gas. Heat combustion is the informational surrogate currently used because it is readily determined. However, this measure relates poorly to waste incinerability. Chlorine and other halogens in the waste tend to extinguish combustion, but simple halogen content give poor indication of incinerability. Autoignition temperature is closely related to incinerability, but for most hazardous compounds, it has not been measured. Better predictors of incinerability could be developed. One scheme, proposed by NBS, would use a combination of factors, but it needs to be tested.
- **There is a lack of basic understanding about how stable toxic PICs are formed.** Some compounds, known to be very difficult to incinerate, also occur as PICs from combusting mixtures of compounds thought to be more easily burned. Our ability to monitor these compounds has only recently made such observations possible, and there are many high-temperature kinetic reactions not fully understood. Unless specifically analyzed, a selected PIC would go undetected. While additional testing of individual combustion facilities will demonstrate specific DRE capabilities, these observations are not likely to improve our fundamental understanding of PIC formation. In particular, with the cost of test burns with POHC monitoring so high, some more basic research on PIC formation would be appropriate. Current EPA research and development, however, is focused in support of near-term permitting activities.

[21]W. Schaub, National Bureau of Standards, personal communication, January 1983.

Review of Selected High-Temperature Treatment Technologies

There are a variety of treatment technologies involving high temperatures which have, or will likely have, important roles in hazardous waste management. Most of these technologies involve combustion, but some are more accurately described as destruction by infrared or ultraviolet radiation.

Discussion below focuses on the distinguishing principles, the reliability and effectiveness, and the current and projected use of these technologies. Unless otherwise noted, DRE values were measured in accordance with EPA testing procedures. Table 32 summarizes the advantages, disadvantages, and status of these technologies.

1. **Liquid injection incineration.**—With liquid injection, freely flowing wastes are atomized by passage through a carefully designed nozzle (see fig. 8). It is important that the droplets are small enough to allow the waste to completely vaporize and go through all the subsequent stages of combustion **while** they reside in the high-temperature zones of the incinerator. Residence times in such incinerators are short, so nozzles especially, as well as other

Table 32.—Comparison of Thermal Treatment Technologies for Hazard Reduction

Advantages of design features	Disadvantages of design features	Status for hazardous waste treatment
Currently available incinerator designs:		
Liquid injection incineration:		
Can be designed to burn a wide range of pumpable waste. Often used in conjunction with other incinerator systems as a secondary afterburner for combustion of volatilized constituents. Hot refractory minimizes cool boundary layer at walls. HCl recovery possible.	Limited to destruction of pumpable waste (viscosity of less than 10,000 SSI). Usually designed to burn specific waste streams. Smaller units sometimes have problems with clogging of injection nozzle.	Estimated that 219 liquid injection incinerators are in service, making this the most widely used incinerator design.
Rotary kilns:		
Can accommodate great variety of waste feeds: solids, sludges, liquids, some bulk waste contained in fiber drums. Rotation of combustion chamber enhances mixing of waste by exposing fresh surfaces for oxidation.	Rotary kilns are expensive. Economy of scale means regional locations, thus, waste must be hauled, increasing spill risks.	Estimated that 42 rotary kilns are in service under interim status. Rotary kiln design is often centerpiece of integrated commercial treatment facilities. First noninterim RCRA permit for a rotary kiln incinerator (IT Corp.) is currently under review.
Cement kilns:		
Attractive for destruction of harder-to-burn waste, due to very high residence times, good mixing, and high temperatures. Alkaline environment neutralizes chlorine.	Burning of chlorinated waste limited by operating requirements, and appears to increase particulate generation. Could require retrofitting of pollution control equipment and of instrumentation for monitoring to bring existing facilities to comparable level. Ash may be hazardous residual.	Cement kilns are currently in use for waste destruction, but exact number is unknown. National kiln capacity is estimated at 41.5 million tonnes/yr. Currently mostly nonhalogenated solvents are burned.
Boilers (usually a liquid injection design):		
Energy value recovery, fuel conservation. Availability on sites of waste generators reduces spill risks during hauling.	Cool gas layer at walls result from heat removal. This constrains design to high-efficiency combustion within the flame zone. Nozzle maintenance and waste feed stability can be critical. Where HCl is recovered, high temperatures must be avoided. (High temperatures are good for DRE.) Metal parts corrode where halogenated waste are burned.	Boilers are currently used for waste disposal. Number of boiler facilities is unknown, quantity of wastes combusted has been roughly estimated at between 17.3 to 20 million tonnes/yr.
Applications of currently available designs:		
Multiple hearth:		
Passage of waste onto progressively hotter hearths can provide for long residence times for sludges. Design provides good fuel efficiency. Able to handle wide variety of sludges.	Tiered hearths usually have some relatively cold spots which inhibit even and complete combustion. Opportunity for some gas to short circuit and escape without adequate residence time. Not suitable for waste streams which produce fusible ash when combusted. Units have high maintenance requirements due to moving parts in high-temperature zone.	Technology is available; widely used for coal and municipal waste combustion.
Fluidized-bed incinerators:		
Turbulence of bed enhances uniform heat transfer and combustion of waste. Mass of bed is large relative to the mass of injected waste.	Limited capacity in service. Large economy of scale.	Estimated that nine fluidized-bed incinerators are in service. Catalytic bed may be developed.
At-sea incineration: shipboard (usually liquid injection incinerator):		
Minimum scrubbing of exhaust gases required by regulations on assumption that ocean water provides sufficient neutralization and dilution. This could provide economic advantages over land-based incineration methods. Also, incineration occurs away from human populations. Shipboard incinerators have greater combustion rates; e.g., 10 tonnes/hr.	Not suitable for waste that are shock sensitive, capable of spontaneous combustion, or chemically or thermally unstable, due to the extra handling and hazard of shipboard environment. Potential for accidental release of waste held in storage (capacities vary from between 4,000 to 8,000 tonnes).	Limited burns of organochlorine and PCB were conducted at sea in mid-1970. PCB test burns conducted by Chemical Waste Management, Inc., in January 1982 are under review by EPA. New ships under construction by At Sea Incineration, Inc.
At-sea incineration: oil drilling platform-based:		
Same as above, except relative stability of platform reduces some of the complexity in designing to accommodate rolling motion of the ship.	Requires development of storage facilities. Potential for accidental release of waste held in storage.	Proposal for platform incinerator currently under review by EPA.

Table 32.—Comparison of Thermal Treatment Technologies for Hazard Reduction—Continued

Advantages of design features	Disadvantages of design features	Status for hazardous waste treatment
Pyrolysis:		
Air pollution control needs minimum: air-starved combustion avoids volatilization of any inorganic compounds. These and heavy metals go into insoluble solid char. Potentially high capacity.	Greater potential for PIC formation. For some wastes produce a tar which is hard to dispose of. Potentially high fuel maintenance cost. Waste-specific designs only.	Commercially available but in limited use.
Emerging thermal treatment technologies:		
Molten salt:		
Molten salts act as catalysts and efficient heat transfer medium. Self-sustaining for some wastes. Reduces energy use and reduces maintenance costs. Units are compact; potentially portable. Minimal air pollution control needs; some combustion products, e.g., ash and acidic gases are retained in the melt.	Commercial-scale applications face potential problems with regeneration or disposal of ash-contaminated salt. Not suitable for high ash wastes. Chamber corrosion can be a problem. Avoiding reaction vessel corrosion may imply tradeoff with DRE.	Technology has been successful at pilot plant scale, and is commercially available.
High-temperature fluid wall:		
Waste is efficiently destroyed as it passes through cylinder and is exposed to radiant heat temperatures of about 4,000° F. Cylinder is electrically heated; heat is transferred to waste through inert gas blanket, which protects cylinder wall. Mobile units possible.	To date, core diameters (3", 6", and 12") and cylinder length (72") limit throughput capacity. Scale-up may be difficult due to thermal stress on core. Potentially high costs for electrical heating.	Other applications tested; e.g., coal gasification, pyrolysis of metal-bearing refuse and hexachlorobenzene. Test burns on toxic gases in December 1982.
Plasma arc:		
Very high energy radiation (at 50,000° F) breaks chemical bonds directly, without series of chemical reactions. Extreme DREs possible, with no or little chance of PICs. Simple operation, very low energy costs, mobile units planned.	Limited throughput. High use of NaOH for scrubbers.	Limited U.S. testing, but commercialization in July 1983 expected. No scale-up needed.
Wet oxidation:		
Applicable to aqueous waste too dilute for incineration and too toxic for biological treatment. Lower temperatures required, and energy released by some wastes can produce self-sustaining reaction. No air emissions.	Not applicable to highly chlorinated organics, and some wastes need further treatment.	Commercially used as pretreatment to biological wastewater treatment plant. Bench-scale studies with catalyst for nonchlorinated organics.
Super critical water:		
Applicable to chlorinated aqueous waste which are too dilute to incinerate. Takes advantage of excellent solvent properties of water above critical point for organic compounds. Injected oxygen decomposes smaller organic molecules to CO_2 and water. No air emissions.	Probable high economy of scale. Energy needs may increase on scale-up.	Bench-scale success (99.99% DRE) for DDT, PCBs, and hexachlorobenzene.

SOURCE: Office of Technology Assessment, compiled from references 12 through 29.

features, must be designed for specified waste stream characteristics such as viscosity. Certain waste must be preheated. Nonclogging nozzles are available, but all nozzles must be carefully maintained. One of the chief costs is maintenance of refractory walls. Incinerator design is a complex, but advanced field. Many distinguishing design features are currently proprietary; especially nozzle designs and refractory composition.

Injection incinerator designs, especially nozzle design, tend to be waste-specific. However, individual designs exist for the destruction of many different liquid waste mixes: motor and industrial oils, emulsions, solvents, lacquers, and organic chemicals of all kinds including relatively hard-to-destroy pesticides and chemical warfare agents.

Liquid injection incinerators, together with rotary kilns (see below) form the current basis of the hazardous waste incineration industry. These technologies have been used for the purpose of destroying industrial waste for many years. In the mid-1970's EPA testing and data reviews of these facilities provided the basis for the current interim performance standard

Figure 8.—Injection Liquid Incineration

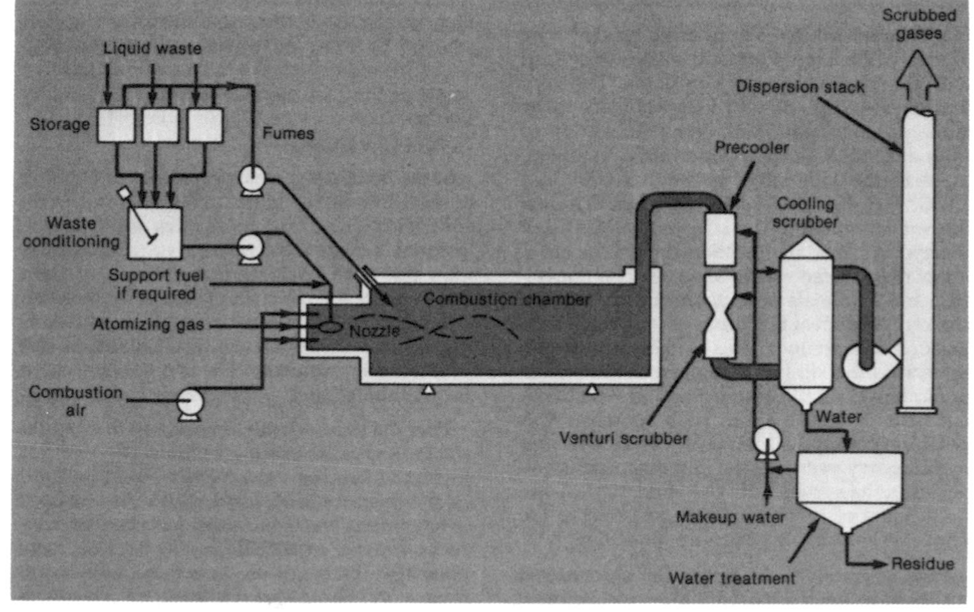

SOURCE: D. A. Hitchcock, "Solid-Waste Disposal: Incineration," *Chemical Engineering*, May 21, 1979.

of 99.99 percent DRE for incineration of hazardous materials.

EPA has recently begun testing incinerators to better understand the DRE capabilities for the most difficult-to-burn waste. Analysis is not yet complete, but preliminary indications both confirm the 99.99 percent capabilities, and underscore the sensitivities of individual incinerators to operational and waste feed variables.[22]

2. **Rotary kilns.**—These can handle a wider physical variety of burnable waste feeds—solids and sludge, as well as free liquids and gases. A rotating cylinder tumbles and uncovers the waste, assuring uniform heat transfer. The cylinders range in size from about 3 ft in diameter by about 8 or 10 ft long, up to 15 or 20 ft in diameter by about 30 ft long. Rotary kilns operate between temperature extremes of approximately 1,500° and 3,000° F, depending on location measured along the kiln. They range in capacity from 1 to 8 tons of waste per hour.[23]

The primary advantage of rotary kilns is their ability to burn waste in any physical form and with a variety of feed mechanisms. Many large companies that use chemicals (such as Dow Chemical Co., 3M Corp., and Eastman Kodak) incinerate onsite with their own rotary kilns. For flexibility, this is often in combination with injection incinerators. Similarly, large waste management service firms (Ensco and Rollins Environmental Services) operate large rotary kilns as part of their integrated treatment cen-

[22]Timothy Oppelt, and various other personal communications, Environmental Protection Agency, Office of Research and Development, IERL, Cincinnati, Ohio, December 1982 and January 1983.

[23]Ibid.

ters. Others are in commercial operation throughout the United States and Europe.[24]

3. **Cement kilns.**—These are a special type of rotary kiln. Liquid organic waste are cofired with the base fuel in the kiln flame. The very thorough mixing and very long residence times make possible more complete combustion of even difficult to burn organic waste. Temperatures in the kiln range between 2,600° and 3,000° F (1,400° and 1,650° C). Also, the alkaline environment in the kiln neutralizes all of the hydrochloric acid produced from the burning of chlorinated waste. Most ash and nonvolatile heavy metals are incorporated into the clinker (product of the kiln) and eventually into the cement product. Heavy metals incorporated into the clinker may present either real or perceived risks (toxicological and structural), but little is known about such concerns.[25] A portion (perhaps 10 percent) of the ash and metals carry over into the kiln dust that is collected in the system's air-pollution control system. Some of this material is recycled to the kiln, the balance is generally landfilled.[26] [27]

Five controlled test burns for chlorinated waste have been documented in wet process cement kilns in Canada, Sweden, Norway, and most recently, the United States. The foreign results have tended to confirm the theoretical predictions—that 99.99 percent or better can be achieved for chlorinated hydrocarbons. However, these studies lack strong documentation of control protocols. In the Swedish results, representative concentrations of very difficult to burn waste were destroyed beyond the limits of monitoring technology, indicating better than 99.99 percent destruction.[28] The California Air Resources Board recently recommended the use of cement kilns to destroy PCB waste.[29] EPA has recently completed a carefully controlled test on the most difficult to burn waste at the San Juan Cement Co. in Duablo, Puerto Rico. The results of this test are still being evaluated.

Some hazardous wastes are currently being burned in cement kilns under the energy recovery exclusion, but, these have been generally nonhalogenated solvents or waste oils, rather than the most toxic and/or difficult to burn compounds, for which they may be well suited. Since 1979 the General Portland Co. of Paulding, Ohio, has been burning 12,500 tons per year of nonhalogenated waste solvents as a supplemental fuel.

There is theoretically no limit on the fuel-to-waste-feed ratio; as long as the waste mix has sufficient heating value, a kiln could be fired solely on waste feed. Idled kilns could be used as hazardous waste facilities. Local public concerns, notably over spills during hauling, have presented the major obstacle to such incinerator use, but commercial interest apparently is still strong.[30] Much will depend on how new regulations affect land disposal use.

4. **Boilers.**—Ignitable waste with sufficient heating value are coincinerated with a primary fuel in some types of boilers. The boiler converts as much as possible of the heat of combustion of the fuel mix into energy used for producing steam. Different types of boilers have been designed to burn different types of fuels. Boilers burn lump coal, pulverized coal, No. 2 oil, No. 6 oil, and natural gas.[31] The predom-

[24] *Technologies for the Treatment and Destruction of Organic Wastes as Alternatives to Land Disposal,* State of California, Air Resources Board, August 1982.

[25] Myron W. Black, "Impact of Use of Waste Fuels Upon Cement Manufacturing," paper presented at the First International Conference on Industrial and Hazardous Wastes, Toronto, Ontario, Canada, October 1982.

[26] Douglas L. Hazelwood and Francis J. Smith, et al., "Assessment of Waste Fuel Use in Cement Kilns," prepared by A. T. Kearney and the Portland Cement Association for the Office of Research and Development, EPA, contract No. 68-03-2586, March 1981.

[27] *Alternatives to the Land Disposal of Hazardous Wastes,* op. cit.

[28] Robert Olexsey, "Alternative Thermal Destruction Processes for Hazardous Wastes," Environmental Protection Agency, Office of Research and Development, May 1982.

[29] "An Air Resources Board Policy Regarding Incineration as an Acceptable Technology for PCB Disposal," State of California, Air Resources Board, December 1981.

[30] Myron W. Black, "Problems in Siting of Hazardous Waste Disposal Facilities—The Peerless Experience," paper presented at a conference on Control of Hazardous Material Spills, 1978.

[31] Environmental Protection Agency, Office of Research and Development, "Technical Overview of the Concept of Disposing of Hazardous Waste in Industrial Boilers," contract No. 68-3-2567, October 1981.

inant application to hazardous waste involves boilers of the kind that would normally burn No. 2 fuel oil.

These boilers are similar to liquid injection incinerators, but there are important differences with respect to the high destruction efficiencies desirable for hazardous waste: 1) they have purposefully cooled walls, and 2) at least some of the walls and other parts exposed to the combustion products are often metallic instead of refractory. The reason that the walls of the boiler must be cooled is to make use of the heating energy from the product gas. In the combustion chamber, this results in a relatively cool area (a thermal boundary layer) through which combustion products might pass.

The metallic surfaces avoid some expensive refractory maintenance but the bare metal surfaces are susceptible to corrosion where halogenated organic waste are burned. For this reason industrial boiler owners, concerned for the life of their equipment, probably limit their use of such waste. However, there is a growing industrial trend toward recovery of hydrochloric acid from the stack gas.* [32] Acid recovery requires that stack gas temperatures greater than 1,200° C be avoided, since this condition shifts the chemical equilibrium toward free chlorine. For hazardous waste destruction, however, higher temperatures are better.

For these reasons, efficient boilers must be designed so that hydrocarbon destruction occurs mostly **in the flame zone** with very little reaction occurring after the flame zone. As is the case with incinerators, boiler design is well advanced, and many designs are proprietary. High fuel efficiency designs may recirculate the flame envelope back into itself to enhance the formation of the series of reactions necessary for complete combustion. Other designs may involve staged injections with varying waste-to-fuel ratios.[33]

*Currently a significant amount, perhaps over 4 percent of the U.S. hydrochloric acid, is produced from stack gas scrubbers. Half of this is from boilers and half from incinerators.

[32]James Karl, Dow Chemical Co., personal communication, January 1983.

[33]Elmer Monroe, DuPont Chemical Co., personal communication, December 1982.

Evaluating the actual hazardous waste destruction capabilities of various boilers has only just begun by EPA. Only three tests were complete at the time of this report; seven more are planned. Tests to date have been conducted primarily with nonhalogenated, high heating value solvents and other nonhalogenated materials. These tests have demonstrated DREs generally in the 99.9 percent area. Subsequent testing will be directed toward waste that are considered to be more difficult to destroy than those tested up to this point.[34] Testing at cooperating boiler facilities is expected to confirm that boilers of a wide variety of sizes and types can achieve hazardous waste destruction efficiencies comparable to those achieved by incineration for some common waste fuels. Industry cooperation will be needed, though, for field testing of those difficult-to-burn and the more toxic wastes marginally useful as fuels.

Actual waste destruction achieved through coincineration probably has more to do with how and why the boiler is operated, and with knowledge of the waste feed contents, than with the type and size of boiler. Destruction by combustion for toxic organic compounds requires very complete, efficient combustion. Thus, in a boiler, the objective of getting usable heat out of the fuel mix is **similar** to that of achieving high destruction of toxic organic waste. **However, the marginal benefits of achieving incremental degrees of destruction may be valued differently by different users.** For example, it may cost less at very large boilers (e.g., those at utilities and large industrial facilities) to save fuel costs through increased combustion efficiency than at smaller boilers. Thus, utility boilers are probably designed and operated for stringent fuel efficiency by an economic motivation that may parallel the rigorous incinerator performance standard in its effect for DREs. Although there would be an economic advantage for these facilities to burn waste fuels, many would not be able to find reliable and sufficient supplies. On the other hand, the objective with many industrial boilers is to **deliver** an **optimal** amount of heat

[34]Olexsey, op. cit.

over time. Thus, achieving 100-percent combustion efficiency is not desirable if it takes 2 days to achieve this goal. **Incinerators have as their direct goal the destruction of the fuel compound which is not so in boilers.**

Excluding the very largest utility and industrial boilers, there are about 40,000 large (10 million to 250 million Btu/hr) industrial boilers and about 800,000 small- to medium-sized institutional, commercial, and industrial boilers nationwide.[35] It is expected that most of the industrial boilers having firing capacities less than 10 million Btu/hr may not readily lend themselves to coincineration.[36]

Finally, there are about 14 million residential, single-home boilers which could burn hazardous waste.[37] These small boilers could have adverse health effects on small, localized areas. **In addition, any fuels blended with organics and illegally burned, in apartment houses or institutional boilers, for example, should be expected to reduce the lives of these boilers through corrosion.**

To assess the role that boilers currently play in hazardous waste management nationwide, it is necessary to know what compounds are being burned, in which facilities, and with what DREs. **Without reporting requirements for coincineration, information is seriously lacking.** Currently, boilers may be burning twice the volume of ignitable hazardous waste that is being incinerated. Except for those from petroleum refining, all were discharged to the environment until environmental, handling, or increasing primary fuel costs encouraged their use as a fuel.[38] Of the entire spectrum of burnable waste, those having the highest Btu content are attracted to boilers. This may have economic effects on regulated incineration, because some hazardous waste incinerators could also benefit from the fuel value of the same waste used as auxiliary fuel in boilers.

5. **Multiple hearth incinerators.**—These use a vertical incinerator cylinder with multiple horizontal cross-sectional floors or levels where waste cascades from the top floor to the next and so on, steadily moving downward as the wastes are burned. These units are used **primarily** for incineration of sludges, particularly those from municipal sewage sludge treatment and, to a much lesser extent, certain specialized industrial sludges of generally a low-hazard nature. They are used almost exclusively at industrial plants incinerating their sludges on their own plant site for the latter cases.[39] Such incinerators are not well suited for most hazardous waste for two reasons: they exhibit relatively cold spots, and the waste is introduced relatively close to the top of the unit. Because hot exhaust gases also exit from the top, there is the potential for certain volatile waste components to short-circuit or "U-turn" near the top of the incinerator and exit to the atmosphere without spending an adequate time in the hot zone to be destroyed. This may be improved by having a separate afterburner chamber, but this option does not appear to have become accepted in the hazardous waste field.[40]

At least one brief test of a typical multiple hearth furnace was conducted in the early 1970's, in which the sewage was "seeded" with a small quantity of pesticide material. Although the pesticide was not detected in the exhaust, the researchers became aware of the short-circulating and residence-time problems and did not pursue the application of multiple hearths to hazardous wastes any.

6. **Fluidized bed combusters.**—This is a relatively new and advanced combuster design being applied in many areas. It achieves rapid and thorough heat transfer to the injected fuel and waste, and combustion occurs rapidly. Air forced up through a perforated plate, maintains

[35] M. Turgeon, Office of Solid Waste, Industrial and Hazardous Waste Division, EPA, personal communication, January 1983.

[36] Olexsey, op. cit.

[37] C. C. Shih and A. M. Takata, TRW, Inc., "Emissions Assessment of Conventional Stational Combustion Systems: Summary Report" prepared for the Office of Research and Development, EPA, September 1981.

[38] EPA "Technical Overview of the Concept of Disposing of Hazardous Wastes in Industrial Boilers," op. cit.

[39] *Alternatives to the Land Disposal of Hazardous Wastes,* op. cit.

[40] Oppelt, op. cit.

a turbulent motion in a bed of very hot inert granules. The granules provide for direct conduction-type heat transfer to the injected waste. These units are compact in design and simple to operate relative to incinerators. Another advantage is that the bed itself acts as a scrubber for certain gases and particulates. Its role in hazardous waste may be limited to small and specialized cases due to difficulties in handling of ash and residuals, low throughput capacity and limited range of applicable waste feeds.[41]

There are presently only about 200 such combusters in the United States, used chiefly for municipal and similar sludges. About nine are used for hazardous waste.[42] Existing fluidized bed combusters are sparsely distributed and relatively small. Future applications of fluidized bed technology to hazardous waste is likely to occur at new facilities built specifically for this purpose rather than at existing municipal facilities.

Recent EPA testing at the Union Chemical Co., Union, Me., is still being evaluated. Early test results are mixed with regard to 99.99 percent destruction.[43] The simplicity of this technology and its ease of operation seem to indicate high reliability for achieving those levels of destruction and wastes for which it will prove to be applicable. A catalytic, lower temperature fluidized bed technology is being developed which may have lower energy costs, and may be more applicable to hazardous waste destruction.[44] However, incompatibilities between catalysts proposed on various hazardous waste may present problems to overcome.

7. **Incineration at sea.**—This is simply incinerator technology used at sea, but without stack gas scrubbers. (The buffering capacity of the sea and sea air is the reason for the lack of a scrubber requirement.) Free from the need to attach scrubbers, marine incinerator designers can maximize combustion efficiency in ways that land-based incinerators cannot.[45] Incinerators based on oil drilling platforms would further be freed from accommodating rolling ship motion.

Various EPA monitoring of test commercial burns of PCBs and government burns of herbicide Agent Orange and mixed organochlorines in the mid and late 1970's confirmed the 99.99+ percent destruction capability for liquid injection incineration used at sea.[46] Current technology exists only for liquids. Rotary kilns could be adapted to ships and more readily to oil drilling platforms.

There exists considerable controversy about the test burns recently conducted for PCBs destruction onboard the M.T. Volcanus. Data results are not yet available. Major concerns are whether the land and marine alternatives represent the same environmental risk and if the performance standards are evenly applied. EPA's view is that they represent roughly the same risk.[47] Regarding the performance standards, it should be recognized that the scrubbing the exhaust gas of land-based incinerators may be providing the fourth nine in their DRE performance. Thus, an at-sea DRE of 99.9 percent may be more similar to the land-based DRE of 99.99 percent than it may appear. The contribution of scrubbers to DRE values are not well known.

Additional concerns about incineration at sea include: stack gas monitoring, which is difficult enough on land and perhaps more so on a ship at sea, and the risk of accidents near shore or at sea. The ecological effects of a spill of Agent Orange on phytoplankton productivity could be substantial.[48] Storage facilities necessary for drilling platform-based incinera-

[41] *Alternatives to the Land Disposal of Hazardous Wastes*, op. cit.
[42] Proctor and Redfern, Ltd., and Weston Designers Consultants, "Generic Process Technologies Studies" (Ontario, Canada: Ontario Waste Management Corp., System Development Project, August 1982).
[43] J. Miliken, Environmental Protection Agency, personal communication, November 1982.
[44] R. Kuhl, Energy Inc., Idaho Falls, Idaho, personal communication, January 1983.
[45] K. Kamlet, National Wildlife Federation, *Ocean Dumping of Industrial Wastes*, B. H. Ketchum, et al. (ed.) (New York: Plenum Publishing Corp., 1981).
[46] D. Oberacker, Office of Research and Development, IERL, Environmental Protection Agency, Cincinnati, Ohio, personal communication, December 1982.
[47] Ibid.
[48] Kamlet, op. cit.

tion may involve still higher spill risks. Public opposition to hazardous waste sites applies also to storage of waste at ports.

Other High Temperature Industrial Processes

Other types of applicable combustion processes including metallurgical furnaces, brick and lime kilns, and glass furnaces, are examples of existing industrial technologies which might be investigated as potential hazardous waste destruction alternatives.[49] There is no reporting of such uses that may be occurring, and no DRE data have been collected. The beneficial use exclusion may apply to many of such practices.[50] However, objectives of such processes are not necessarily complementary or supportive of high DRE. The technical potential for hazardous waste destruction and need for regulation of such practice needs investigation.

Emerging Thermal Destruction Technologies

Undue importance should not be placed on the distinction between current and emerging technologies. The intent is merely to distinguish between technologies currently "on the shelf" and those less commercially developed for hazardous waste applications.

Pyrolysis.—This occurs in an oxygen deficit atmosphere, generally at temperatures from 1,000° to 1,700° F. Pyrolysis facilities consist of two stages: a pyrolyzing chamber, and a fume incinerator. The latter is needed to combust the volatilized organics and carbon monoxide produced from the preceding air-starved combustion. The fume incinerator operates at 1,800° to 3,000° F. The pyrolytic air-starved combustion avoids volatilization of any inorganic components and provides that inorganics, including any heavy metals, are formed into an insoluble easily handled solid char residue. Thus, air pollution control needs are minimized.[51]

[49] PEDCO Environmental Services, Inc., "Feasibility of Destroying Hazardous Wastes in High Temperature Industrial Processes," for the Office of Research and Development, IERL, EPA, Cincinnati, Ohio, May 1982.

[50] 40 CFR, sec. 261 (c)(2).

[51] *Alternatives to the Land Disposal of Hazardous Wastes*, op. cit.

Pyrolysis has been used by the Federal Government to destroy chemical warfare agents and kepone-laden sludge and by the private sector to dispose of rubber scrap, pharmaceutical bio-sludge, and organic chlorine sludges. Most recently, pilot plant test burns on chlorinated solvents from a metal-cleaning plant have been destroyed with 99.99 percent destruction.[52]

Broader application would await much more equipment development and testing. Among the potential problems with pyrolysis are:

- Greater potential for toxic and refractory PICs formation than with combustion in air. The reducing atmosphere produces larger amounts of these compounds, and they may pass through the off-gas afterburner.
- Production of an aqueous tar that may be difficult to dispose in either a landfill or an incinerator.
- Substantial quantities of auxiliary fuel may be required to sustain temperature in the afterburner.[53]

Commercially, high throughput (up to 6,500 lb/hr) and required air pollution control requirements may be key future benefits. However, maintenance costs due to moving parts, and the need for well-trained operators may be relatively high.[54]

Molten Salt Reactors.—These achieve rapid heating and thorough mixing of the waste in a fluid heat-conducting medium. Liquid, solid, or gaseous wastes are fed into a molten bath of salts (sodium carbonate or calcium carbonate). Solids must be sized to 1/4- or 1/8-inch pieces in order to be fed into the bed. The bed must be initially preheated to 1,500° to 1,80C° F. Provided that the waste feed has a heating value of at least 4,000 Btu/lb, the heat from combustion maintains the bed temperature, and the combustion reactions occur with near completion in the bed instead of beyond it. The sodium carbonate in the bed affects neutraliza-

[52] Oppelt, op. cit.

[53] Ibid.

[54] *Technologies of the Treatment and Destruction of Organic Wastes as Alternatives to Land Disposal*, op. cit.

tion of hydrogen chloride and scrubbing of the product gases. Thus, the bed is responsible for decomposition of the waste, removal of the waste residual, and some off-gas scrubbing. A bag house for particulates completes air pollution control and the removal system.[55] (See fig. 9.)

In EPA tests, a pilot scale unit (200 lb/hr) destroyed hexachlorobenzene with DRE's exceeding 6 to 8-9's (99.9999-percent to 99.999999-percent destruction and removal) and chlordane with DREs exceeding 6 to 7-9's.[56] Rockwell International also claims 99.999-percent destruction efficiencies* from private tests on malathion and trichloroethane.

Reactor vessel corrosion has impeded development of molten salt destruction (MSD). Vessel corrosion is accelerated by temperature, reducing conditions (less than sufficient oxygen), and the presence of sulfur. Traditionally, MSD reaction vessels have been refractory lined, presenting operational and maintenance costs similar to those of conventional incinerators. Rockwell International offers an MSD system with a proprietary steel alloy reactor vessel. This vessel is warranted for 1 year if the system is operated within specified ranges of temperature, excess air, and melt sulfur content.[57]

Ash as well as metal, phosphorous, halogen, and arsenic salts build up in the bed and must be removed. In the case of highly chlorinated waste (50 percent or more) the rate at which salt must be removed approaches the rate of waste feed. Both the salt replacement (or regeneration) and residual disposal determine economic viability for a given application. In pro-

[55] Ibid.

[56] S. Y. Yosim, et al., Energy Systems Group, Rockwell International, "Molten Salt Destruction of PCB and Chlordane," EPA contract No. 68-03-3014, Task 21, final draft, January 1983.

*Not DRE; small amounts removed in bed salts and baghouse treatment were not measured.

[57] J. Johanson, Rockwell International, Inc., personal communication, January 1983.

Figure 9.—Molten Salt Destruction: Process Diagram

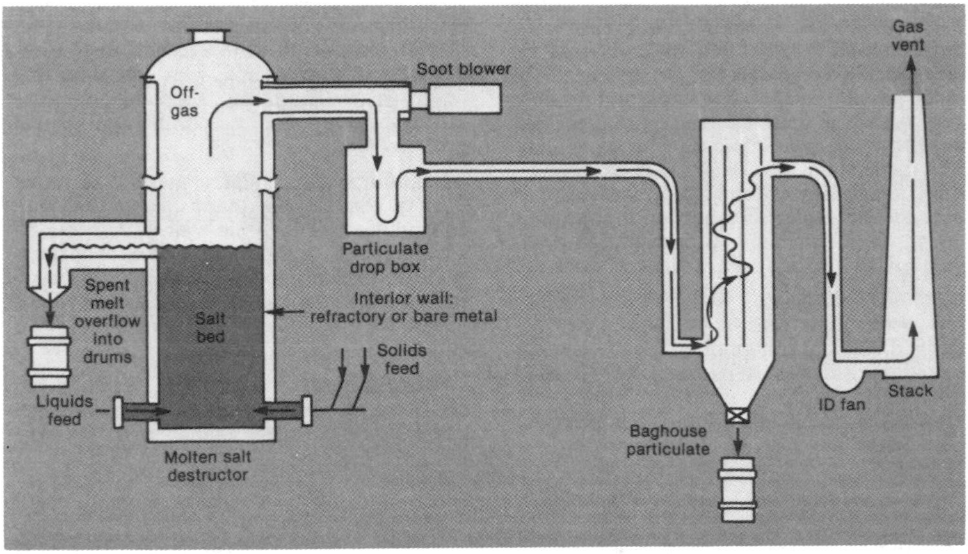

SOURCE: Adapted from S. Y. Yosim, et al., Energy Systems Group, Rockwell International, "Molten Salt Destruction of HCB and Chlordane," EPA contract No. 68-03-3014, Task 21, final draft, January 1983.

posed commercial ventures, sodium chloride residue would be landfilled and calcium chloride would be sold as road salt or injected deep well.[58]

The process is intended to compete with rotary kilns and may find application for a broad market of wastes that are too dilute to incinerate economically. However, water in the waste feed, as with any incineration technology, uses up energy in evaporation. Due to the extremely high DREs demonstrated in pilot scale tests, the process is expected to be very attractive for destroying the highly toxic organic mixtures and chemical warfare agents, which currently present serious disposal problems. Rockwell International is in final negotiations with two commercial ventures in California and Canada. Commercial-scale units offered are 225 and 2,000 lb/hour.[59]

High Temperature Fluid-Wall Reactors.—In these reactors, energy is transferred to the waste by radiation (rather than by conduction and convection as in the above processes). A porous central cylinder is protected from thermal or chemical destruction by a layer of inert gas. The gas is transparent to radiation, and the cylinder is heated by radiation from surrounding electrodes to 3,000° to 4,000° F. The refractory cylinder reradiates this energy internally to the passing waste.[60] The important result is very rapid and thorough heating of the waste stream for complete combustion or generation. The speed of the heating presents little opportunity for the formation of intermediate products for incomplete combustion that present concerns in conventional incineration processes. Also, process control is good since the radiation is directly driven by electricity.

A bench-scale reactor (¼ lb/min) has destroyed PCBs in contaminated soil (1 percent by weight) with 99.9999 percent DRE.[61] In addition, the Thagard Research Corp., which conducted the tests, claims that it has privately burned hexachlorobenzene with 99.9999 percent DRE in a 10 ton per day unit.[62] A commercial-scale unit (20 to 50 tons per day) is operated as a production unit by a licensee in Texas, which has agreed to allow Thagard to continue hazardous waste destruction demonstration burns there.[63] In December 1982, California and EPA conducted demonstration burns of some gases that are difficult to destroy thermally—1,1,1-trichloroethane, carbon tetrachloride, dimethyl chloride, Freon 12®, and hexachlorobenzene. Results are currently being assessed.

Further scaleup may be needed to provide commercial throughput, and this will involve larger ceramic cores. The effects of thermal stresses on the life of the cores present the major untested concern for scale up.

Near-term commercialization of the Thagard reactor is planned. During 1983, a Miami investment firm is expected to underwrite the development of a mobile reactor, reducing breakdown and setup time from several weeks to only a few days. This will facilitate the collection of test burn performance at potential applications sites.[64] Also, Southern California Edison Inc. is considering the process for future destruction of PCB-laden soil and for stabilization of a variety of its heavy metal-bearing liquid waste. The utility is also interested in selling byproducts of carbon black from the process.[65]

In addition to its potential mobility resulting from its compact design the only air pollution control need for the fluid wall reactor may be a bag house to control particulates. The process is not expected to be economically competitive with conventional incineration, but will be applicable especially to contaminated soils and silts.

Plasma-Arc Reactors.—These use very high energy free electrons to break bonds between molecules. A plasma is an ionized gas (an elec-

[58]Ibid.
[59]Ibid.
[60]*Technologies for the Treatment and Destruction of Organic Wastes as Alternatives to Land Disposal,* op. cit.
[61]E. Matovitch, Thagard Research Corp., personal communication, January 1983.
[62]Ibid.
[63]Ibid.
[64]Ibid.
[65]E. Faeder, Southern California Edison Power Co., personal communication, January 1983.

trically conductive gas consisting of charged and neutral particles). Temperatures in the plasma are in excess of 50,000° F—any gaseous organic compounds exposed to plasma are almost instantly destroyed. Plasma arc, when applied to waste disposal, can be considered to be an energy conversion and transfer device. The electrical energy input is transformed into a plasma. As the activated components of the plasma decay, their energy is transferred to waste materials exposed to the plasma. The wastes are then atomized, ionized, and finally destroyed as they interact with the decaying plasma species. There is less opportunity for the formation of toxic PICs. Most of the destruction occurs without progression of reactions which could form them.[66]

Private tests conducted for the Canadian Government have demonstrated 99.9999999 percent (i.e., 9-9's) destruction on pure transformer fluid (58 percent chlorine by weight).[67] Depending on the waste, the gas produced has a significant fuel value.[68] A high degree of process control and operational simplicity are additional advantages. For halogenated waste (a major market target), the gases would have to be scrubbed but the scrubbers needed are very small.

The process is in the public domain and nearing commercialization. The developer plans to market mostly small, self-contained, mobile units. Costs are intended to be competitive with incineration.[69] The first commercial application is planned to be in operation in July 1983.

Wet Oxidation.—Proven in commercial application, wet oxidation processes can destroy reliably nonhalogenated organic waste (e.g., cyanides, phenols, mercaptans, and nonhalogenated pesticides). The oxidation reactions are fundamentally the same as in combustion but occur in liquid state. Since it is not necessary to add large quantities of air as in incineration, potentially contaminated gas emissions are avoided. The reactions take place at temperatures of 430° to 660° F (and pressures of 1,000 to 2,000 psi). For many applicable waste feeds, the oxidation reaction resulting produces enough heat to sustain the process, or even to produce low pressure steam as an energy by-product. The oxidation reactions typically achieve 80 percent complete decomposition to carbon dioxide and water, and partial decomposition to low molecular weight organic acids of the remaining waste feed.[70] Currently, the process remains commercially applicable to aqueous organic waste streams which are too dilute for incineration, yet too toxic for biological treatment.

Still in development are catalytic modifications to the wet oxidation process, aimed at the more stable highly chlorinated organics. Bench-scale tests conducted by I. T. Enviroscience have demonstrated that a bromide-nitrate catalyst promotes completeness of oxidation. Should this process achieve destructions similar to those of incineration, its lack of air emissions, and the ease of using performance monitoring would be advantageous.[71]

Super Critical Water.—At temperatures and pressures greater than 374° C and 218 atm, water becomes an excellent solvent for organic compounds and can break large organic molecules down into molecules of low molecular weight.[72] In a system patented by Modar, Inc., injected oxygen completely oxidizes the lower molecular weight molecules to carbon dioxide and water. DDT, PCBs and hexachlorbenzene have been destroyed with efficiencies exceeding 99.99 percent in bench-scale testing.[73] Costs are expected to be highly dependent on scale.[74] If high-destruction efficiency is maintained

[66]C. C. Lee, Office of Research and Development, IERL, Environmental Protection Agency, personal communication, January 1983.

[67]Plasma Research Inc., unpublished test results, January 1983.

[68]*Alternatives to the Land Disposal of Hazardous Wastes*, op. cit.

[69]T. Barton, Plasma Research Inc., personal communication, January 1983.

[70]P. Shaefer, Zimpro, Inc., personal communication, November 1982.

[71]Oppelt, op. cit.

[72]M. Modell, "Destruction of Hazardous Waste Using Supercritical Water," paper delivered at the 8th Annual Research Symposium on Land Disposal, Incineration, and Treatment of Hazardous Wastes (Fort Mitchell, Ky.: Environmental Protection Agency, Mar. 8, 1982).

[73]Ibid.

[74]*Alternatives to the Land Disposal of Hazardous Wastes*, op. cit.

through scaleup, this could be an attractive alternative to incineration.

Biological Treatment

Conventional biological treatments use naturally occurring organisms to degrade or remove hazardous constituents. In contrast, biotechnology uses bacteria which have been selected from nature, acclimated to particular substrates, and mutated through methods such as exposure to ultraviolet light for fixation of the adapted characteristics. Many toxic substances cannot be degraded biologically, although they may be effectively removed from a waste stream this way. Types of conventional biological techniques, waste stream applications, and their limitations are listed in table 29. These techniques have found widespread use for treatment of municipal and industrial wastes to prevent the formation of odorous gases, to destroy infectious micro-organisms, to remove nutrients for aquatic flora, and to remove or destroy some toxic compounds. Several biological techniques may be used as a series of steps to treat a waste, including ending with landfarming (also called land spreading or land treatment). The latter refers to the deposit of a waste, or some sludge or residue from a treatment, onto land or injected some small distance beneath the surface. Naturally occurring organisms in the soil degrade the waste, usually organic, and periodic plowing may be necessary to ensure adequate oxygen levels for degradation.

The physical, chemical, or biological processes that can be used to eliminate or reduce the hazardous attributes of wastes exist in as many forms as those processes used to manufacture the original material. All of these treatments produce waste residuals; usually a liquid and a solid waste. The hazardous characteristics of these waste residuals must be evaluated in terms of the objective desired for their final disposition or recovery. Without such an objective it is difficult to evaluate the benefit, either economic or environmental, of applying the treatment process. These treatments *can* result in merely changing the form or location of the waste. For example, concentrating organics from a dilute waste stream does not necessarily provide any benefit in terms of increased protection of health. If this separation and concentration treatment allows the waste constituent to be recovered or, alternatively, makes a destruction technology viable, the treatment has been beneficial.

Residuals from hazardous waste treatments are discharged to surface waters, to publicly owned wastewater treatment works (POTWs) or are sent to landfills or land treatment disposal. To the extent that the treatments considered below can reduce the toxic characteristics of wastes through destructive or degradative reactions, they are similar in their effect to thermal destruction technologies. To the extent that they are able to mitigate specific hazard characteristics, they render the wastes nonhazardous. And, to the extent that they reduce the mobility of the waste, they reduce the interaction of land-disposed wastes with the environment.

Many references exist describing unit physical, chemical, and biological processes and how they may be combined. This discussion will not attempt to duplicate any such descriptive listings. Table 30 lists the established applications. Selection of one or several processes depends on such factors as waste feed concentration, desired output concentration, the effects of other components in the feed, throughput capacity, costs, and specific treatment objectives.

Appendix B
Landfills of the Future: Aboveground and Aboveboard

KIRK W. BROWN

Landfills have come a long way from the days of open dumping and ravine filling. New hazardous waste landfills are required to have leachate collection systems, liners, and caps. If properly constructed and maintained, these measures should substantially reduce the short-term quantity of leachate that escapes from a landfill. Liners, leachate collection systems, and caps on below ground landfills may, however, cease to function properly sometime within the first few decades after closure of the landfill. There is now no requirement for leachate collection following the 30-year closure period. Yet many wastes are likely to remain hazardous for centuries.

A failed liner or leachate collection system may not be detected and, if detected, may be impossible to repair without taking drastic measures such as waste excavation. Many of these problems could be avoided by constructing future landfills aboveground and on a sloped, double-lined base (Figure 1A). The expense of building a sloped base could be avoided by constructing a "hillfill" using gently sloping hills as the landfill base. Much of the technology already developed for existing landfills, including double liners, leachate collection systems, waste stabilization, and caps could be readily adapted to aboveground landfills.

LEACHATE COLLECTION SYSTEMS

Below ground landfills do not have leachate collection systems designed for continuous leachate removal. Most of these systems cannot remove

This article originally appeared in *Pollution Engineering* in November 1983. Reprinted by permission.

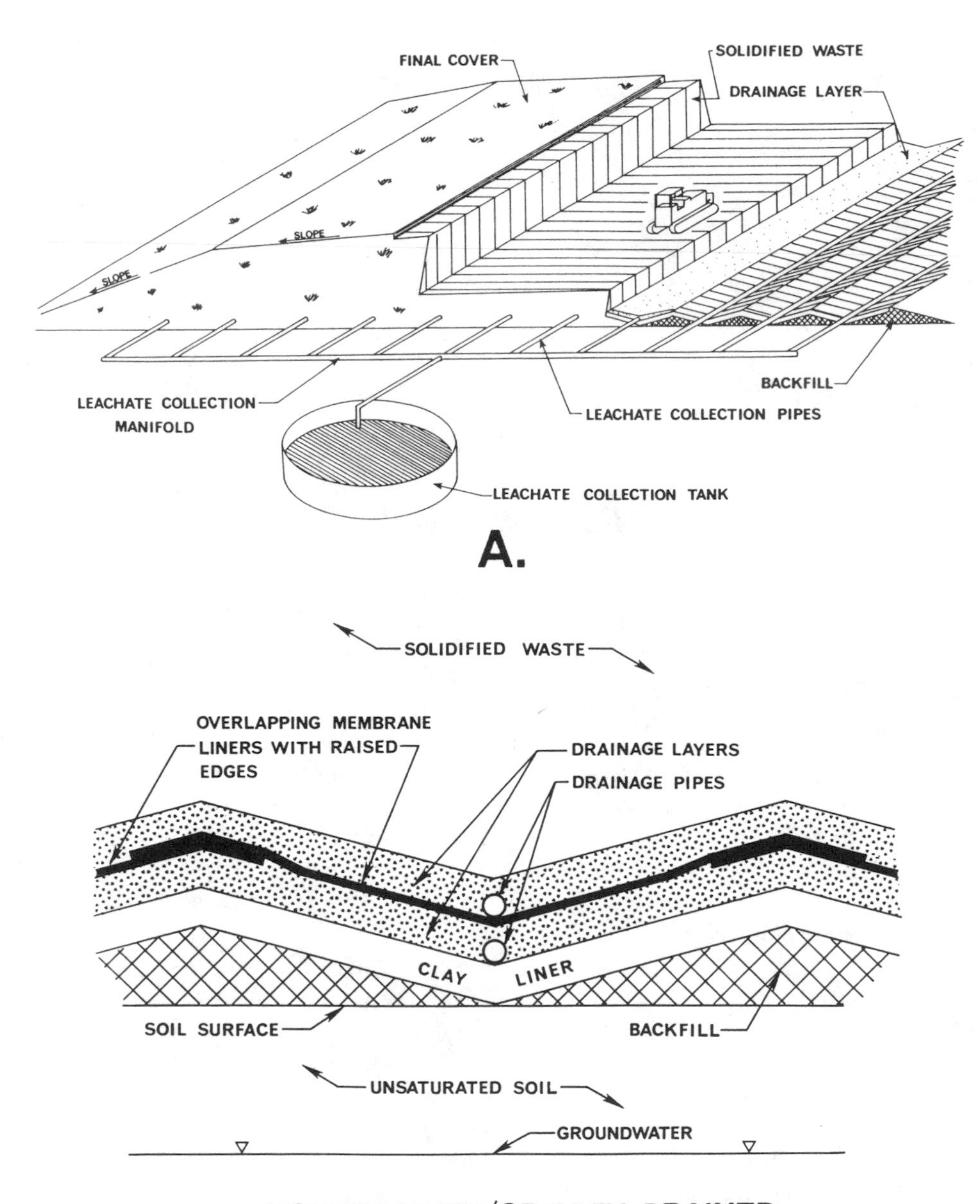

Figure 1. (A) Landfills should be constructed on an aboveground sloped base to facilitate any needed repairs; **(B)** the leachate collection system over the sloped base should include both drainage layers and drainage pipes so that either drainage system would continue to remove leachate from the surface of the liners even if the other system failed.

leachate until the collection pipes are submerged in it. Consequently, below ground landfills may well have at least shallow pools of leachate standing on their liners at all times. Futhermore, no provisions are currently being made for monitoring or removing leachate after the sites are completely closed. Thus, leachate is likely to accumulate over time. If a leachate collection system fails due to clogging or collapse of the collection pipes, it is extremely difficult to repair.

An aboveground landfill may be constructed with a continuously operating leachate collection system. By simply incorporating a slope in the base of an aboveground landfill, leachate would be continuously removed by gravity. Since all leachate collection pipes would be above ground level, they would be accessible, in the event that repairs are needed, and serviceable for centuries. The system would also be relatively inexpensive because no pumps or labor would be required for leachate removal.

With both a drainage layer and collection pipes constructed over the sloped liner (Figure 1A), leachate would continue to be removed by gravity even if the collection pipes collapsed. Consequently, there would be little opportunity for other problems to occur, such as seepage through either the liner or sidewalls.

LINERS AND CAPS

Flexible membranes and clay are the two main materials used to construct low permeability landfill liners. Clay liners may be rendered significantly more permeable by exposure to concentrated leachates that hazardous waste may initially release. While flexible membrane liners may have initial low permeabilities, their 10 to 30 year useful lifetime covers only a fraction of the period during which a landfill may generate contaminated leachate. Any hazardous waste disposal facility would be better off with a double liner system, and the best double liner would be a combination of a flexible membrane upper liner with an underlying clay liner (Figure 1B). Compatibility tests may be used to select the best membrane material for containing particular waste leachates.

A properly installed and tested membrane liner should last at least ten years if the need to rely on the seams could be eliminated. This would be possible if flexible membrane liners were constructed on a sloped, sawtoothed base as shown in Figure 1B. Because leachate would be continuously removed by gravity, the liquid level should never reach the overlapped membrane edges. By the time the membrane deteriorated, the concentration of salts, acids, bases, and organics should have decreased substantially.

Clay minerals, with their proven ability to last thousands of years, may be a better liner material for minimizing the long-term leakage of weak leachates. With an overlying membrane liner, the clay liner would be protected from the strong initial leachate that might otherwise increase liner permeability.

With continuous removal of leachate, liners in aboveground landfills should have less exposure to leachate and should, therefore, leak much less and last much longer than similar liner configurations in below ground landfills. Consequently, there would be less dependence on site geology for containment of the waste and a lower potential for groundwater contamination. These and other advantages of landfilling waste using an aboveground sloped liner are presented in the following list:

1. No head of liquid would collect on the liner due to continuous gravitational drainage. Consequently:
   a. Potential for groundwater contamination would be minimized.
   b. Dependence on site geology for waste containment would be lessened.
   c. Potential would be reduced for liner deterioration from contact with leachate.
2. A leachate collection system constructed over the aboveground sloped base and consisting of drainage layers and perforated collection pipes would provide for the following:
   a. Leachate collection pipes would be accessible for cleaning out should they become clogged.
   b. Continuous gravitational removal of leachate would continue through the drainage layer even if the collection pipes completely collapsed.
   c. Any increase in the volume of leachate produced would be an immediate indication of the need for cap repair.
   d. Long after closure, if leachate were released, it would be readily observable rather than move undetected into groundwater.

Both above and below ground landfills may be equipped with low permeability caps. While caps reduce leachate generation resulting from infiltration of water, these caps may shear or crack due to settlement of the landfilled waste. Compared to leaking liners, cracked caps may be relatively easy to repair. However, cap deterioration may not be readily apparent through visual inspection once the landfill is covered with topsoil and permanent vegetation. Cap failure may only be detected through increase in leachate production. Such an increase in leachate volume may go unnoticed in a below ground facility until the groundwater was polluted. However, it would be readily evident as an increased leachate discharge rate in an aboveground landfill.

ADVANTAGES AND DISADVANTAGES OF ABOVEGROUND LANDFILLS

Continuous leachate removal and the resultant increased effectiveness of liners in aboveground landfills may greatly reduce the potential for both groundwater pollution and long-term liability. Since aboveground facilities would be further removed from groundwater, siting requirements would be less severe than those for below ground facilities. With gases generated above the ground surface, the possibility of subsurface migration to adjacent areas would be virtually eliminated. Advantageous options that could be incorporated into aboveground landfills include the following:

1. Above the leachate collection system, layers of materials such as crushed limestone and activated charcoal could be placed to remove heavy metals and organics, respectively.
2. Before placement of the low permeability cap, leachate could be recirculated through the landfill cells to both hasten digestion of readily degradable organics and leach the highly mobile waste constituents prior to the post-closure period.

There are a variety of other advantages to keeping waste aboveground. For instance, no one would ever forget the location of the disposal site. In addition, if the waste became valuable someday, it could easily be mined. The uncapped landfill surface could be used as an intensive land treatment unit.

Disadvantages of aboveground landfills include the potential for poor site aesthetics and the need for erosion control. However, with proper design and use of vegetation, both problems could be overcome. Site aesthetics of aboveground facilities may be improved by maintaining a permanent vegetative cover over the facility. Additional improvements in site aesthetics could be obtained by planting trees and small woody species around the periphery and along adjacent roadways. Erosion could be minimized by a permanent vegetative cover and a site design that minimized steepness and total area of side slopes. Using adjacent landfill cells would decrease the total area of slopes, while a small amount of earth-fill could sufficiently reduce the steepness of exposed sidewalls.

ADVANTAGES AND DISADVANTAGES OF BELOW GROUND LANDFILLS

The main advantage of below ground landfills is that these facilities are out of sight and hence out of mind. The main disadvantage is that these facilities eventually get out of control. One technique used to site below ground landfills in humid climates is to artificially lower the ground-

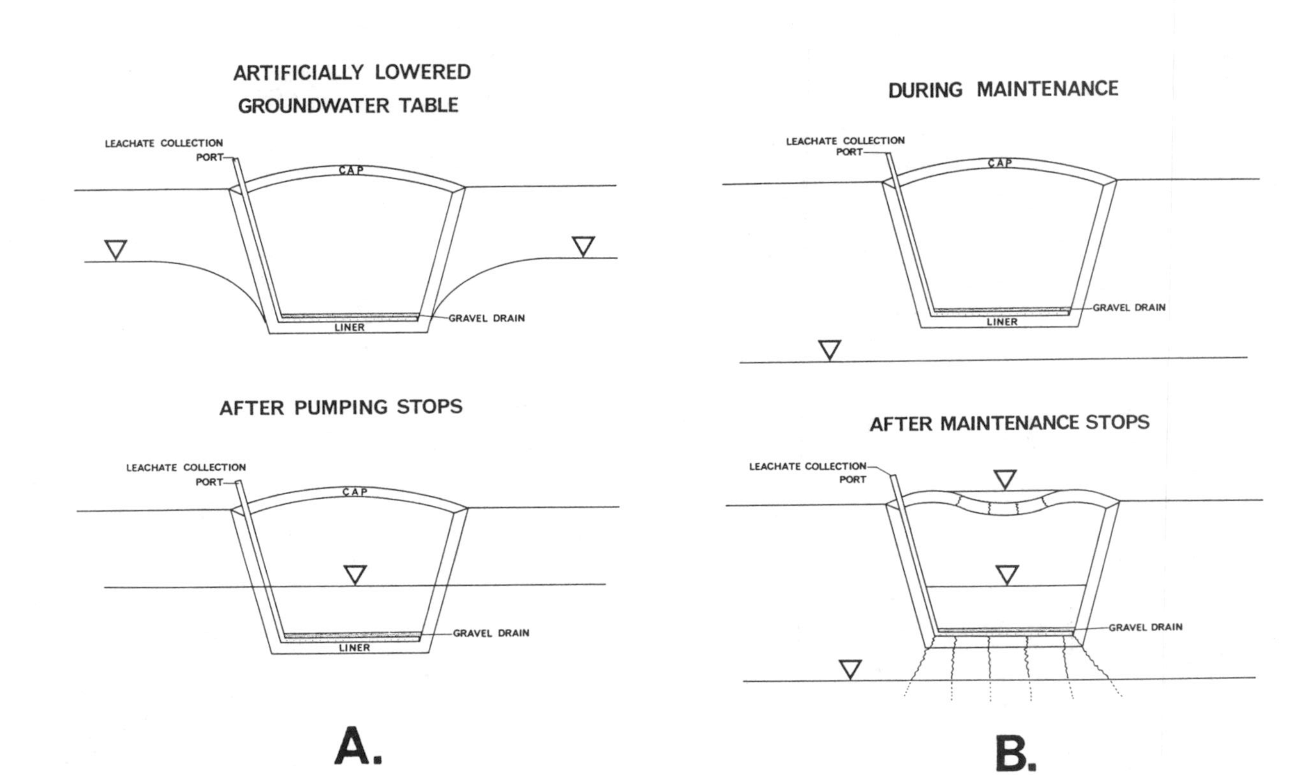

Figure 2. **(A)** Below ground landfills sited within a groundwater table require perpetual pumping to prevent groundwater contamination; **(B)** State-of-the-art below ground landfills may only reduce the potential for groundwater contamination until maintenance stops.

water table by continuously pumping out leachate (Figure 2A). This technique is supposed to force liquids to flow into the landfill, thereby preventing contamination of groundwater or deterioration of the liner due to exposure to leachate. The main problem with this design is the unrealistic requirement for perpetual pumping.

State-of-the-art below ground landfills minimize groundwater contamination by pumping out leachate and maintaining a low permeability cap (Figure 2B). While these measures should prevent widespread groundwater contamination during the active life of a facility, the situation after maintenance stops may be quite different. As organic wastes decompose, voids form in the fill which may well cause the cap to settle and crack. Liquids may even pool on the cap or move through the cap and pool on the liner. The hydraulic gradient formed due to the pooled leachate could then accelerate leakage, liner deterioration, and groundwater contamination.

IMPROVING LANDFILLS

Landfills have come a long way in the past two decades, but there is still need for improvement. Lack of public acceptance and the threat of long-term liabilities due to groundwater contamination from below ground landfills are two of the indications that further changes are warranted. Current hazardous waste regulations limit the landfilling of liquids. The next step should be to reduce the landfilling of organic chemicals by encouraging the use of land treatment (for readily degradable organics such as oily refinery wastes) and incineration (for non-degradable organics such as PCBs). The remaining inorganic wastes and incinerator residues could be solidified and safely disposed of in aboveground landfills.

Well-designed aboveground facilities have the potential to both improve the acceptability and greatly decrease the pollution risks of landfills. Landfilling waste over an aboveground sloped base is an idea whose time has come.

Appendix C
Same Wastes, New Solutions: The Market for Treatment Alternatives

RICHARD C. FORTUNA

INTRODUCTION

Over seven years have passed since the original enactment of the Resource Conservation and Recovery Act (RCRA), the nation's basic authority for preventive management of hazardous wastes. While some progress has been made in establishing a partial program, from the standpoint of ultimate treatment many of the implementing regulations have done little more than legally sanction practices that are the environmental equivalent of illegal dumping. Moreover, the program has failed to regulate meaningfully or restrict the practice that is the leading cause of this nation's hazardous waste problem—land disposal. As a result, the nation has little more to show in the way of a preventive hazardous waste program than it did at the time of RCRA's initial passage.

The intervening years from 1976 to the present have taught us far more about the nature, scope, and severity of the problem than they have witnessed solutions implemented to address them. The fact that the Congress initially allowed only 18 months for the Environmental Protection Agency to promulgate all necessary hazardous waste regulations is perhaps the most telling indicator of our matured comprehension of the problem's magnitude. At the same time, it is abundantly clear that we can afford to delay no longer the imposition of meaningful restrictions and prohibitions on the land disposal of hazardous wastes if the treatment goals of the program are to be realized, if treatment is to become the primary method of waste management, and if we are to halt the creation

This appendix is based upon a presentation before the Bureau of National Affairs Conference on "The New EPA" in Washington, D.C., on September 26, 1983.

of additional Superfund sites due to unsound present practices. While the goals of the Act and its regulatory program have always acknowledged the desirability of treatment, they have failed to provide any explicit directives or substantive means toward that end.

The Hazardous Waste Treatment Council believes that there is an overriding need for substantive change in national policy, a change that explicitly restricts and prohibits certain land disposal practices and provides the regulatory tools to accomplish this task. In fact, the most important element of the formula for change in this nation's hazardous waste management practices is one which is not currently within the sole control of the Agency, irrespective of leadership; rather, it lies with the Congress.

The following discussion will examine the reasons why treatment has not been employed as the primary method of waste management, and the necessary national policy remedies to make treatment a national reality.

THE CAPABILITY AND CAPACITY TO BEGIN THE TRANSITION

Prior to discussing the reasons behind the status quo and the rationale for change, it is important to note that there are few if any wastes that cannot be treated in lieu of or prior to land disposal. Beyond capability, the nation also is fortunate to have a respectable base of existing capacity to begin the transition away from land disposal and toward the primary use of ultimate treatment.

A recent National Academy of Sciences report concluded that there are no current waste streams that cannot be treated by existing technologies either separately or in combination (National Academy of Sciences, Committee on Disposal of Industrial Hazardous Wastes, February, 1983). The report's conclusions comport with the Council's views, and, moreover, put to rest the common misconception that the transition to treatment methods must somehow be predicated upon a high-tech revolution in the distant future. To be certain, new and more efficient methods of treatment are being developed, but rather than await a high-tech revolution we must revolutionize our national policy so as to utilize existing basic technology that has been demonstrated and which is presently available. The perfect must not be allowed to become the enemy of the good.

Regarding the current national capacity to manage wastes in alternative ways, there is no doubt that additional permitted capacity will be needed to cope with the growing restrictions on halogenated organics, corrosives and certain metal-containing wastes that are the subject of actions at the state and national level. However, the majority of the treatment firms that the Council has surveyed reported unused capacity in 1982 between 30-50 percent. As of late 1983, this trend still holds with the possible exception being those facilities that are permitted to incinerate PCBs. This current high rate of utilization, however, is largely attributable to a liquid PCB storage deadline under the Toxic Substances Control Act (TSCA) that is unlikely to sustain such demand for more than a few months. In the event that the Congress rectifies the numerous deficiencies in the national program in the manner that will be discussed subsequently, the Council believes that the nation can insure substantial expansion of the commercial (off-site) treatment market for the full range of hazardous wastes. In fact, the Council has recently surveyed its membership and related treatment firms on the likely expansion of existing capacity through the year 1986 in the event that RCRA policies are rectified to restrict land disposal and in turn require treatment as the primary method of waste management. While the results of this first-of-its-kind survey are not yet final, it is clear that the nation can expect at least a doubling of its current commercial capacity to treat halogenated organic and other hazardous wastes.

These figures and projections are indeed significant, for while only 5 percent of all hazardous waste generated in the country is managed at commercial facilities, over 80 percent of all generators rely on commercial facilities to manage at least half of their wastes (Preliminary Findings of National Survey of Hazardous Waste Generators, conducted for the Office of Solid Waste by Westat Inc., 8/30/83). In addition, commercial facilities will continue to manage the wastes that are uneconomic and difficult to treat.

The Council also views treatment capacity availability on a national, or at the very least, regional basis. Just as one does not find a refinery at every drilling site, so, too, treatment facilities depend upon economies of scale that virtually always require the interstate transportation of hazardous wastes. In fact, with higher overhead and initial capital costs, economies of scale are of even greater significance to treatment facilities.

If there is a base of existing methods and capacity, and the commercial industry is poised to expand to meet the expected demand, why has it taken so long and what is required to bring about the beginning of this long-overdue transition to treatment? These questions will be examined in the proceeding sections.

IMPEDIMENTS TO THE USE OF TREATMENT AND CONTINUED INCENTIVES FOR DISPOSAL

The Market for Treatment and the State of Regulatory Coverage

There are few who would challenge the need for increased use of high-technology hazardous waste treatment. Moreover, based on the actions of several states, few could question whether such a transition is underway. This movement is premised on the belief that treatment, while not being magical in and of itself, when properly conducted provides certainty in two key respects: certainty in knowing what was done to the wastes, and certainty in knowing that future generations will not be exposed to their hazards. While these are compelling and self-evident benefits of treatment, there are equally telling reasons why treatment has not emerged to any significant degree.

First, and foremost, the bottom line of treatment and disposal is the bottom line. Hazardous wastes are like water running down hill; they will always be disposed of along the path of least regulatory control and least cost. The market alone is an unreliable and indifferent broker when it comes to insuring protective management of hazardous wastes in a cost competitive environment. The forces that establish the lowest marketplace cost also underwrite the methods which provide the least protection for public health and the environment when practiced without restriction. This view shares such wide acknowledgment that even the spokesperson for the nation's largest landfill firm has affirmed the axiomatic correlation between waste disposal practices and cost (*Wall Street Journal*, 6/10/83).

Second, all forms of waste management must play by the same rules, with the common denominator being equal protection of public health and the environment, and not simply the lowest cost. Only through a consistent and comprehensive national policy and implementing regulations can the nation insure that the drive for the lowest bottom line does not in turn provide an unacceptable minimum level of public health protection. As a recent EPA report concluded, "the Federal hazardous waste statutes in combination do not cover many of the major sources and types of hazardous wastes." ("Evaluation of Market and Legal Mechanisms for Promoting Control of Hazardous Wastes," Industrial Economics Inc., prepared for Office of Solid Waste, pursuant to Executive Order 12291, September, 1982).

Present policies must do more than merely "encourage" or create "incentives" for treatment. The disparity in the use of land disposal over treatment is founded in a concomitant disparity between the regulatory requirements for the respective operations which in turn allows cost to be the only factor dictating management choice. These are not solely the observations of the Treatment Council. As one industry analyist recently stated when speaking of the underlying principles by which generators make decisions regarding waste disposal methods, "there is an enormous incentive to dispose of waste cheaply" (*Wall Street Journal*, 6/10/83).

Treatment and Disposal—A Study in Contrasts

The specific reasons for low cost unprotective land disposal are numerous: little capital is required up front; there are no inherent limitations on what can be placed into a land disposal facility; many preventive measures such as dual liners and groundwater monitoring are either not required or avoided due to weak regulations and "grandfathering"; ultimate liability for the facility after closure may be shifted to governmental entities; and protection of public health is predicated solely upon physical barriers that cannot contain many wastes that are so disposed, rather than specifically restricting the types of wastes that can be land disposed. No matter how heroic the engineering effort, there is a wide range of wastes that cannot be contained by land disposal facilities. As such, technical standards alone are a necessary though insufficient means of controlling the release of hazardous wastes from land disposal facilities (Montague, P., Princeton University, September, 1982; 46 FR 11128, 2/5/81). This latter recognition has been the single most important factor in crystallizing a concensus on the need for policies that affirmatively bring about the use of primary treatment.

While the regulations and the inherent nature of the practice place no restrictions on land disposal, a treatment facility invests the majority of its capital before a single load of waste is ever received. In addition, most forms of treatment operate under stringent standards governing process efficiency and duration and are specific to given waste streams. Unlike landfills, there is no current treatment process that can treat all types of wastes. Most importantly, it will always cost more to render permanently a waste non-hazardous at the time of generation under controlled conditions than it will simply to bury or inject and hope. The desire for increased certainty in the protection of public health cannot be separated from the inevitability of increased costs for proper hazardous waste treatment.

While treatment delivers greater certainty that wastes will be prevented from causing future threats to public health and the environment, certainty is a two-way street. The regulations and policies must provide greater certainty that there will be a market for something other than unrestricted land disposal. Without an explicit policy that requires treatment as the primary method of waste management, the envisioned transition is little more than a pipe dream. As an article in the *Wall Street Journal* (8/15/83, p. 19) observed, "There is a lot of risk involved in designing and siting a waste treatment plant. Why should a company take a technological risk along with other risks?"

That same article also observed that treatment methods will not be employed as long as there are so many legal ways to dispose of wastes that are not subject to control, when certain facilities are not required to meet public health standards, and when there are no qualitative or quantitative restrictions as to waste placement on the land. It is the overriding view within the treatment community that the market for alternatives cannot be separated from the adequacy of the regulatory system under which we operate. The inadequacies or loopholes in the present system are the single most important contributor to the improper disposal of wastes and to the undermining of fair competition in the hazardous waste management industry. The loopholes alone are more significant from the standpoint of treatment than the combined effect of illegal dumping and the lax enforcement of existing regulations.

In summary more generators and individual waste streams are exempt from RCRA coverage than are subject to present regulations. Moreover many hazardous waste facilities remain totally exempt. This is particularly true for "recycling" facilities such as industrial boilers, furnaces, and kilns that can burn hazardous wastes for the purpose of energy recovery without regard to emissions limitations or destruction efficiency.

While numerous examples of such loopholes have been previously documented in Congressional hearings and in this text, new ones continue to be discovered. Most notable among these are the cases of ethylene dibromide (EDB) and deep-well injection. The EPA recently imposed a virtually unprecedented suspension of certain pesticide uses of EDB. Nevertheless, the wastes from the production of this very pesticide are neither listed nor identified as hazardous wastes under RCRA. The same situation exists for most species of dioxin-containing wastes. Regarding deep-well injection, it is now clear that more hazardous wastes are managed by this method than all others combined, approximately 10 billion gallons per year. At the same time, deep wells are not required to carry third-party liability coverage like other RCRA facilities, are difficult if not impossible to monitor beyond the injection stack, and frequently inject corrosive and organic wastes that are both incompatible with and erode the injection formation (Sullivan, P.J.,"An Assessment of Class I Hazardous Waste Injection Wells," Ball State University, prepared for the EPA Office of Research and Development pursuant to the AAAS/EPA Environmental Sciences and Engineering Fellows Program, 1983). Despite these serious fundamental uncertainties and deficiencies with deep-well injection, the report also notes that Regions IV and V expect a doubling of deep-well injection activity by 1986. Not surprisingly, deep-well injection is frequently less expensive than even direct land burial of the waste.

In light of the continued proliferation of threats to public health if practices are not changed, and given that the emergence of treatment practices requires more specific knowledge of the market sector that will be dedicated to treatment prior to investment, the policy changes that will bring about the use of treatment cannot be left to chance or allowed to occur by happenstance. As with every other environmental statute, unless compliance is required for all types of facilities and generators and within the same general timeframe, few will want to be the first to step forward and incur a competitive disadvantage. Indeed, without a predictable market for alternative methods, the risks of financial failure and the fear of competitive disadvantage will outweigh any general notion of where the system should be headed and the status quo will persist.

The following examples vividly illustrate the manner in which wastes are being shifted to the environmental medium of least control, and the relationship between regulations, cost, and the conditions that will bring about protection of public health through treatment rather than disposal:

- The Agency has documented (48 FR 14489) where a generator accumulated emissions control dust from steel production (hazardous waste K061) for over eight years without recycling, properly treating, or disposing of the waste. At present, over 40,000 tons are now piled in the open in an abandoned quarry near a drinking water source.
- The recent release of the "Surface Impoundment Assessment" (Office of Drinking Water, July, 1983) clearly shows that the overwhelming preponderance of the nation's surface impoundments either leak or are likely to. Fully 39 percent of the 8,000 active impoundments used for the storage and disposal of wastes was determined to have a high potential to contaminate groundwater. Included in the impoundment study are those used in treating wastewaters for the purpose of complying with the requirements of Clean Water Act discharge limitations. It is sobering and at the same time ironic to learn that the cost of protecting one enviornmental medium, surface water, is the contamination of another medium, groundwater; a medium where contamination is tantamount to irrevocable destruction of the resource. As a recent study concluded, "prevention of groundwater contamination is a more effective strategy than cure" (*Science*, 8/19/83, 713).
- The much discussed prohibition on landfill disposal of containerized liquids has done little more than spur the use of kitty litter. For example, rather than taking a barrel of methlene chloride to be treated, it is instead mixed with kitty litter to remove free-standing liquids and then landfilled, increasing the total cost by little more than $5 per drum from what land disposal otherwise would have cost and with no greater protection of the public health.
- To add insult to injury, there are no restrictions whatsoever on the bulk disposal of liquid wastes into landfills and other land disposal facilities, which accounts for far more waste than containerized liquid disposal. The restrictions on containerized liquids do not apply to bulk because codisposal of liquid hazardous wastes with solid waste is not considered to be a form of bulk landfill disposal. This particular perversity has had a significant adverse effect on the use of treatment. For example, operators of inorganic liquid treatment facilities have seen the costs of land disposal at certain competing facilities drop by almost 70 percent, thereby maintaining an untouchable cost advantage over the treatment facility. This example is particularly telling because the landfill involved practices codisposal, merely spreading wastes over trash without proper treatment. Moreover, the treatment facilities involved were initially cost competitive with the landfill, separated by only pennies per gallon for the respective processes. When the lack of substantive requirements or preventive measures allows such

slack in the cost structure, it is easy to see how the majority of wastes continues to be land disposed.

- Regarding the initial cost to construct a treatment versus disposal facility, in one situation a treatment facility operator has expended over one million dollars on blueprints alone to develop a state-of-the-art facility. For far less exposure, the operator could have designed and installed several no-tech surface impoundments for solar evaporation and still have been within the regulations.
- Both the threshold decision to choose treatment over disposal and the decision on the scale of the facility are totally dependent upon the policies which insure that wastes must be managed in a protective manner. For example, if a landfill is built with either too much or too little capacity, it will last a little bit longer or a little bit less than projected. If, however, a treatment plant is improperly scaled because of an overestimate of waste volume, it may well spell financial ruin for that facility.

A POLICY OF PROHIBITIONS

Presumptive Prohibitions and Statutory Directives

In 1976 the Congress and the nation recognized in a very general sense the need for national minimum standards governing the management of hazardous wastes to protect public health and to prevent the creation of waste havens in neighboring states. By 1983, the information base had truly matured, and our understanding of the scope and severity of the problem we face had advanced to the point of defining the true parameters of the problem. With the necessary changes in policy the nation can embark upon the implementation of a program that will bring about ultimate, rather than deferred, solutions to the management of hazardous wastes.

Unless and until wastes are presumptively prohibited by statute and until a conscious national policy is enacted that places treatment at the top of the hazardous wastes management hierarchy, treatment on any significant scale will not occur. Until all forms of hazardous waste management are forced to play by the same rules—be it incineration and burning in boilers, metal fixation and land disposal, thereby providing equivalent levels of public health and environmental protection—we will continue to witness little more than a warmed-over version of the status quo.

The institution of this policy not only calls for the closing of current regulatory loopholes, but more importantly, requires a change in the structure by which these issues are approached, and regulations subsequently promulgated. To rely merely on the Agency's ability to sort through the maze of potential problems without specific guidance and presumption is a de facto invitation to failure. Rather, what is needed is a structured and specific presumption against unrestricted land disposal, with prohibitions on those wastes that are the most frequent and predictable causes of groundwater contamination, such as halogenated organics, cyanides, and several types of corrosive and inorganic metal wastes. The Agency should have the discretion to make distinctions for certain waste under given management conditions, but the presumption must be clear, and the burden must be shifted. Should the Agency fail to make the necessary decision on variances within a specified period of time, the prohibitions would take effect for the entire category except where alternative treatment capacity does not exist.

It has been argued that this approach may punish the disposer in the event that the Agency fails to act on variances for legitimate circumstances in a timely manner. In this context it is frequently cited that the EPA has not met one single regulatory deadline imposed upon it.

While it is true that the Agency has not met a single statutory deadline under the existing regulatory structure, there is ample evidence that an approach which presumptively prohibits wastes by statutory directive not only will succeed, but that it is the only manner in which to proceed if results are to be forthcoming. In fact, isn't the undisputable record of regulatory delay and dalliance in itself the very reason to try a different approach? Do we continue to employ procedures that insure delay, or do we approach the problem from a different perspective which molds the accrued experience of the past seven years into a presumption against those activities that are the leading cause of groundwater contamination? Do we allow another deadline to go by and force the Agency to be sued by those who seek protection of the public health? Do we continue a regulatory procedure that underwrites the incentive for people to sit back and benefit from prolonged delay in the promulgation of regulations? Do we force the Agency to go through a formal proceeding for each and every waste stream to be restricted irrespective of waste similarities, or do we institute a categorical rule of general applicability based upon the evidence of past "Superfund" experience and state regulatory actions and place the burden on the disposer to provide a compelling demonstration regarding a potential variance from such a general rule? Lastly, can the nation afford not to act and thereby continue the presumption that land disposal can be made safe for all wastes?

The Treatment Council believes that the answers to these questions are clear. If we are agreed that a meaningful approach to and restrictions on land disposal are the highest priority of the national hazardous waste program, if we indeed want decisions made in a timely manner, and if we want to insure, rather than hope, that treatment becomes the primary method of waste management, then we are bound to do something other than trust regulatory proceedings and statutory directives that are at best nebulous and naive.

The reasons for the projected success of this alternative approach are based on the experience with similar regulatory situations under RCRA and the Toxic Substances Control Act (TSCA). Three case studies are particularly instructive.

In the 1980 RCRA amendments, the Agency was given an unstructured, open-ended directive to conduct studies on five categories of "special wastes," that would not be listed or identified as hazardous wastes until the completion of these studies and subsequent determinations by the Agency. The merits of these exemptions withstanding, purely as a structure for decision making, the provision has been an abysmal failure. Few of the studies have been completed, some have not even been funded, and no decisions have been made. As a result, these wastes remain exempt. Some of the affected industries have even begun to push for funding of these studies to make nationally consistent and applicable decisions under RCRA. As it turns out, several states have taken to regulating many of these materials, all using slightly different mechanisms and approaches. Despite the specific naming of the wastes, the provision failed to include a time deadline, it lacked a presumption in favor of waste listing, and most importantly, it had no sanction in the event that the Agency did not act. In short, it allowed the Agency to do nothing, and that is precisely what has happened: nothing.

Turning to another aspect of the RCRA program, listing and delisting of hazardous wastes, one sees that the Agency has been substantially more successful at deregulating hazardous wastes than bringing new wastes into the system. Over 250 permanent and temporary delistings have been granted since May 1980 without one new waste being listed, and without any statutory directive to delist wastes. The case here is not against delisting per se; this is a normal part of the process of regulatory review and refinement. Rather, these data strongly suggest that the process has been responsive in those instances where the regulated community has a direct stake in correcting the application of a general rule to their specific situation.

While there has been only one previous statutory prohibition of a waste material, the experience from this one case study is instructive in several respects. In 1976, the same year of RCRA's enactment, the Toxic Substances Control Act prohibited the manufacture and directly governed the disposal of polychlorinated biphenyls (PCBs). The implementation has not been perfect by any means. PCB transportation is not manifested, storage facilities are not permitted, and many PCBs in "totally enclosed uses" are not being removed from the market. However, PCBs are largely being disposed of in a proper and ultimate manner. The regulations promulgated pursuant to this prohibition in 1979 require that liquid PCBs must be incinerated to a destruction efficiency of 99.9999 percent, that is, for every million gallons of PCB material incinerated, only one gallon can go uncombusted. This statutory directive and its regulations not only have brought about increased protection of public health and the environment but also have been the most important factors behind the emergence of a domestic incineration market for PCBs and other hazardous wastes. Without these types of prohibition on PCBs, it is fair to say that domestic incineration capacity would be a minor fraction of that which is presently available. Without similar prohibitions on other waste streams, one can reasonably predict how much of non-PCB hazardous wastes will be ultimately destroyed and treated. Does anyone believe that PCBs would be incinerated to 99.9999 percent destruction efficiency if it were not a regulatory requirement? Does anyone doubt that PCBs would still be on the regulatory back burner if it were not for a clear and unequivocal Congressional directive? In addition, since 1979, several other thermal and chemical destruction methods have emerged that have substantially increased the available capacity for PCB treatment and decreased its cost in the process.

Even in the state of California, which has instituted one of the most comprehensive land disposal prohibition programs, there was a recognition that a simple general directive without presumption or constraint would fail. It was for this reason that a general presumption against land disposal was instituted. There was no need to recreate the wheel for every situation, but it was necessary to allow the general presumption to be overturned in specific situations where compelling demonstrations could be made.

A De Facto National Policy Already in Place

In many respects, the move to restrict land disposal goes beyond considerations of public health alone. In essence, land disposal must be restricted if this method is to be saved. By this I mean that the citizens of this country have already established a de facto policy that there will be few if any new land disposal facilities in the country. Landfills in particular have been and continue to be viewed with such disfavor because they remain as the repositories for all things large and small, liquid and solid. If land disposal facilities were restricted to take only non-liquid wastes and pretreated wastes and residues, then the conditions that created leakage and migration would likely not occur. We will always need landfills for those wastes that truly cannot be treated and for the residues of treatment operations. In addition, simply as a matter of sound planning in an industrial society, we cannot wait until the last landfill has closed before contemplating treatment alternatives. With restrictions and prohibitions in place, the society can virtually assure itself of doubling the existing life of most land disposal facilities.

CONCLUSIONS

We stand at a watershed point in the development and evolution of the national hazardous waste program. Seven years of experience have revealed the full extent and nature of the problem and have demonstrated the deficiencies of previous legislative and regulatory approaches to the problem.

The nation is long past due for a conscious and comprehensive program of waste-specific restrictions on all forms of land disposal to protect future generations from the hazards and uncertainties associated with land disposal, and at the same time to insure that the public has enough confidence in the nature of the practices in order to accommodate those land disposal facilities that will be required.

The problems before us are 99 percent political and 1 percent technological. There are few wastes that cannot be treated, and the capacity now exists to begin a phased, yet deliberate, transition to the primary use of methods that permanently protect public health and the environment. The legislative decisions made in the months ahead will determine whether we are on the road to achieving this objective, or whether four years from now we will continue to bemoan the lack of progress. Toward this end, the members of the Hazardous Waste Treatment Council welcome the end of the beginning for the national hazardous waste program, and in turn the beginning of the end for unrestricted land disposal.

Annotated Bibliography

Alternatives to the Land Disposal of Hazardous Wastes: An Assessment for California. Prepared by The Toxic Waste Assessment Group, Governor's Office of Appropriate Technology, 1981. 288 pp.

This report supplies a step-by-step, clearly articulated reconstruction of California's decisions to restrict land disposal. After charting the generation of wastes per region, reviewing land disposal practices in the state, and itemizing the state's high-priority wastes, the study then provides a useful overview of alternative technologies.

A tough choice is presented. At the time of writing, California was spending $17 million a year to dispose of its high-priority wastes in landfills. Their surveys indicated that this would increase to about $30 million by 1983 in an effort to meet new federal standards. But the state found, in the course of assembling this report, that using alternative treatment technology for California's high-priority wastes would cost only $50 million a year–with the additional $20 to $30 million annual cost spread among 4,000 businesses with gross sales exceeding $30 billion.

The state decided to discourage land disposal, and chapters 6 through 10 report the conclusions and recommendations of that decision. Chapter 6 provides a survey of available technologies for California and major high-priority waste streams: pesticide wastes, PCBs, cyanide wastes, toxic metal wastes, halogenated organics, and non-halogenated volatile organics. Chapter 8 reports on the commercial applications of these technologies, and chapter 9 explains how the state encourages the construction of these new treatment facilities. A 40-page appendix supplies descriptions of the technologies appropriate for any user desiring a shift from dumping.

For a copy, write: Publications and Information, Dept. of Health, 1600 Ninth Street, Sacramento, CA 95814; or call 916–323–6578.

Detoxication of Hazardous Waste. Ed. Jurgen H. Exner. Ann Arbor, Michigan: Ann Arbor Science, 1982. 362 pp. Reviewed by John Rembetski.

Safe hazardous waste disposal has been labeled by environmental experts as *the* environmental concern of the 1980s. The problem crosses the borders of several professional disciplines; engineers, toxicologists, lawyers and politicians are just a few who play an active role in effecting treatment.

Detoxication of Hazardous Waste contains a collection of various approaches to tackling the menace of hazardous waste. Most of the disposal options considered can be grouped under the broad headings of either recovery, treatment, landfilling, or process modification.

Recovery, when possible, is recommended in the introduction as the most desirable detoxication method. The appeal stems from the notion of obtaining economic and environmental value from hazardous waste. Chemical detoxication (for example, dechlorination) and incineration are both prime examples of treatment options and, as such, receive prominent emphasis in this book. Waste solidification is the only process considered that uses landfilling as an integral part of disposal. Process modification to reduce waste stream volume is not covered.

The waste problem is first examined by evaluating the role of incineration as viable treatment. Incineration wins approval for detoxication of non-recoverable wastes, since energy value still may be utilized. Thorough coverage of the technical, legal, and social aspects of incineration is included.

Polychlorinated biphenyls (PCBs) undoubtedly constitute one of the largest chemical hazards in existence today. Their high persistence and resistance to decay only compound their demonstrated toxicity. The full section devoted to PCB detoxication offers an adequate view of the many ways to treat these substances. Consistent with the recovery ethic, dechlorination shines through as an important process that allows resource recovery with a minimal threat of toxic by-products. Incineration appears as a viable PCB disposal option, given sufficient verification of legal guidelines.

The discovery of 7 kilograms of stored dioxin (TCDD) during a hexachlorophene plant takeover near Verona, Missouri, posed a major toxic peril. This scenario lends itself to a useful case study analysis of dioxin disposal management. The decision of Syntex, Inc., to detoxify the dioxin-laden plant led to the choice of photolytic degradation as the treatment. Announcing the achievement of 99.94 percent destruction efficiency, this case study awards the process a "model detoxication" honor.

Biological treatment may be earning its keep as a useful weapon in the detoxication arsenal. The genetic engineering of biological systems is seen to promise increased manipulation of the many physical and biological constraints that plague microbial treatment. At present, biological treatment appears as most useful in dilute wastewater processing after primary treatment. The need for more research is emphasized in this area.

Hazardous waste disposal clearly presents itself as a multifaceted problem. *Detoxication of Hazardous Waste* demonstrates that the technical component is not necessarily the major constraint. This suggests that the foundation is set for some solutions.

Hazardous Waste in America. By Samuel S. Epstein, Lester O. Brown, and Carl Pope. San Francisco, CA: Sierra Club Books, 1982. 593 pp. Charts, notes, bibliography, and extensive appendices.

The American public lives in the shadow of Love Canal. The ever-widening orbit of leaking landfills dims our understanding of the steps we need to take to prevent future contamination and arrest past mistakes. *Hazardous Waste in America,* a considered and considerable encyclopedia, spotlights the extent of America's predicament. With six chapters itemizing the major modes of toxic waste mismanagement, this book alerts us to the pervasiveness of the problem: Love Canal was not an isolated mistake but only one brief instance of a devastatingly recurrent pattern of inexpert decision-making. Three additional chapters on legal means for addressing the problem explain in lay terms the regulatory and legislative reforms required to improve waste management in the 1980s. The book also contains a citizen's legal guide to hazardous waste issues.

The authors are well equipped to fullfill their tasks. Samuel Epstein, professor of occupational and environmental medicine at the University of Illinois, is an expert witness who has authored hundreds of scientific studies on carcinogenic and toxic hazards. Lester O. Brown is a staffer for congressional oversight subcommittees. Carl Pope, an associate director of the Sierra Club, possesses bureaucratic tact and an excited respect for administrative complexity.

The authors compile an alarming, six-page synopsis of the Reagan administration's efforts to deregulate toxic waste controls (pp. 248–53). This is a compressed and telling summary of federal backsliding. One comes to recognize the distinct benefits of coauthorship: written with the urgency, missionary zeal, and shared vision of *The Federalist Papers*, this team effort might be termed "the environmentalist papers." Instead of analyzing the fateful mission of experimenting with free government in a new world, this new set of federalists analyzes the stressful mandate of administering federal controls in a time of sustained recession, exhaustive deregulation, and mounting contamination.

The comparison breaks down when it comes to the mood of the book. While the founding federalists spoke of the role of a central government in an energized, celebratory, and hopeful way, Epstein, Brown, and Pope are unremittingly pessimistic. In at least a dozen passages, these articulate specialists claim that it may prove impossible to arrest toxic contamination in America. "But in spite of all this activity," the authors note after summarizing the complex regulatory apparatus built by Congress and the courts since 1978, "government performance after Love Canal raises serious doubts about its competence to resolve this problem with conventional regulatory approaches. Indeed, there is reason to wonder if government can even successfully regulate the more egregious abuses found in the disposal of toxic materials" (p. 199).

This two-sentence concession of measured fatalism encapsulates both the strengths and shortcomings of this book. *Hazardous Waste in America* is useful when it informs its readers about mismanagement and failures in command-and-control regulations. They are right: the volume of toxic wastes exempt from federal regulations exceeds the poisons currently controlled. This book examines the most flagrant exemptions and explains why these long-standing loopholes are conspicuously bad. In short, this new Sierra book is the best collection we have, since Michael Brown's *Laying Waste* (1979), for alerting the public to the need for regulatory reform.

Despite its critical value, the book offers little ground for hope. Herein lies its limitation. The authors prefer to describe case after case of mismanagement rather than to undertake the formidable task of analyzing corrective options. The three concluding chapters on technical solutions, future strategies, and present alternatives to dumping are weak and abbreviated. Each ignores a 15-year legacy, left by European industrial giants like France and West Germany, concerning ways to treat, rather than dump, the bulk of toxic wastes. This concluding section also ignores progress made by states, such as California, in shifting from dumping to detoxification technologies and initiating waste reduction strategies.

Moreover, there is a shocking absence of detail concerning corporations in this book, except to recite old incriminating evidence. In fact, the book fails to mention some remarkable good news: a number of industrial trade groups—including the Hazardous Waste Treatment Council and the National Solid Waste Management Association—have begun to fight for stronger regulations in an effort to stabilize the American treatment market. Although the authors admit that "the attempt to solve the problems of hazardous waste disposal by *regulation alone* is doomed to failure" (p. 254), they fail to examine viable management options expressed by dozens of state reports and diverse ten-year development plans published by the waste treatment community. In short, *Hazardous Waste in America* succeeds as an updating of the problem since Love Canal, but ignores the managers, engineers, policymakers, and concerned citizens who have begun to steer the country through this critical challenge of our day.

Laying Waste: The Poisoning of America by Toxic Chemicals. By Michael Brown. New York: Pantheon Books, 1979. 363 pp.

In breaking the story about Love Canal, Michael Brown received three nominations for a Pulitzer for this forcefully written book. Since it first appeared in 1979, much has happened regarding toxic waste problems. Yet it is still the most comprehensive and readable introductory account of the toxic waste crisis. Assembling case after case of contamination and reconstructing the reasons for delays in remedial action, the book exposes the recurrent patterns of mismanagement that created Love Canal and related incidents.

Michael Brown is a crisis journalist; the burden of his news is inherently emotional. Each chapter describes alarming incidents: 800 pounds of chemical residues from industrial lagoons wash into Lake Michigan after each heavy rainfall; hundreds of drums of lead, mercury, arsenic, and nitroglycerine, illegally stacked three deep, exploded in a toxic waste dump in Elizabeth, New Jersey; underground toxic vapors are vented alongside occupied trailers in Anaheim, California.

It requires a temporary repression of outrage, in reading Brown's work, to realize the pervasiveness of this problem. Initially, his work imparts a feeling of helplessness if not the paranoia of sustained surprise. This may be its single weakness. Although each chapter presents its evidence accurately and convincingly, his work is distracted, at times, by exotic and bizarre facts: a pond containing PCBs, benzene, and DDT was found not far from home plate at Shea Stadium: Brown can't resist announcing such distressing news. The uninformed reader is disarmed by such evidence. But once the initial shock of these alarming details wears off, a larger set of questions emerge, many of which have shaped the nature of the debate since.

This is the best book to read for an introduction to the problem. Don't look for an analysis of corrective options in this book; but instead, be prepared for an elaborate and moving account of the origins, developments, and lasting power of the problems.

Love Canal: Science, Politics and People. By Adeline Gordon Levine. Lexington, Massachusetts: Lexington Books, 1982. 263 pp.

The disaster at Love Canal is obviously a subject that deserves serious study, and it seems unlikely that it ever will get more intimate and detailed treatment than in this new book. A professor of sociology at the State University of New York at Buffalo, Adeline Levine spent three years interviewing hundreds of residents at Love Canal, as well as dozens of the state task force members who eventually evacuated the site. In addition, she assembled evidence from several thousand newspaper items, reports, letters, and related documents.

Levine tells a dual story: the day-to-day tale by which residents of the contaminated zone became convinced that they shared a common source of their ailments, and then the more arduous tale by which local, state, and federal officials reluctantly accepted the evidence of contamination. Both parties were suffering from an inherited trust that the invisible threat of toxic contamination was mostly bogus. At times, neighbors suffering the exact same symptoms failed to share their troubles. Throughout, blue-ribbon scientific panels and expert chromosome-damage researchers underestimated the tenacity of the problem.

Levine's seventh chapter is her strongest. Here she evaluates the shared beliefs of the Love Canal Homeowners' Association, explaining the means by which the residents appropriated information about the puzzling and frightening turn in their lives. If she had resorted more frequently to an analysis of beliefs in this manner, the book would have proven more telling.

Profit from Pollution Prevention. By Monica Campbell and William Glenn. Toronto, Ontario: Pollution Probe Foundation, 1982. 416 pp., 130 illustrations, 75 tables, 140 photographs.

This guide to industrial waste reduction and recycling is appropriate for American readers on a number of counts. First, it documents hundreds of success stories of large and small firms that have turned wastes into assets. By providing brief profiles of these innovative businesses, the larger argument shows how companies have saved money through product reformulation, process redesign, more efficient equipment installation, recycling, and inter-industry waste exchanges.

The first chapter provides, in detail, a successful waste reduction strategy. The bulk of the book contains industry-specific chapters that detail waste sources and pollution prevention opportunities in the following areas: electroplating; dry cleaning; printing; ink formation; paints and coatings manufacture; coatings application; solvent-using industries, such as painting, printing, and adhesives industries; waste-oil-generating industries, including automotive shops and general manufacturing operations; industries that generate waste sulphur; industries that generate waste fly ash; plastics processing and reprocessing; textiles manufacture and finishing, including dyeing; food processing and by-product utilization; photography and photofinishing; tanning; pulp and paper manufacture; and many industries that generate exchangeable waste streams.

Each topic area includes an extensive "for further information" section which identifies relevant associations, journals and references for follow-up investigation, both in the U.S. and abroad. Additional chapters discuss inter-industry waste exchanges, waste recovery technologies, and waste treatment and disposal technologies. For the layperson, a comprehensive glossary of technical terms is included. Further, an index pinpoints innovative technologies by both type and company name. Although directed primarily at the business community, this book can serve government officials, management consultants, and general readers interested in moving America beyond dumping. Simply put, the efficacy of our neighbor's innovative approaches serves as a model of what is possible and necessary here in the U.S.

For other work on the Canadian response to hazardous waste problems, see *Alternatives: A Journal of Friends of the Earth, Canada.* The Fall-Winter 1982 double issue specializes on toxic wastes, printing 11 essays on legal, industrial, and political issues. (Vol. 10, nos. 2 and 3.) Both the book and the magazine are available from Friends of the Earth, P.O. Box 569, Station B, Ottawa, Canada K1P 5P7.

Siting Hazardous Waste Facilities: Local Opposition and the Myth of Preemption. By David Morell and Christopher Magorian. Cambridge, Massachusetts: Ballinger Publishing Company, 1982. 226 pp.

Siting of new hazardous waste facilities is the responsibility of the states and their component local governments. But as the public becomes more and more traumatized by the mention of toxic wastes, the establishment of safe, locally acceptable facilities has become one of the most difficult aspects of hazardous waste management.

The above title is the most useful general study of this siting problem. Morell and Magorian explain the considerable impediments that retard the construction of safe facilities in America. Much of the intense local opposition that has thwarted siting results from an uneven distribution of costs and benefits: large per capita losses are concentrated in a particular locality or region in order to provide diffuse benefits for the entire state. The authors argue that siting thereby requires a redistribution of benefits and a careful balancing of costs. The trick is to make the siting process open, sequential, and timely so that it is perceived as legitimate and necessary by developers, residents, and state authorities. The authors explain this process effectively, with an informed sense of the urgency of the problem and the political prerequisites for a prompt solution.

Technologies and Management Strategies for Hazardous Waste Control. By the Office of Technology Assessment. Washington, DC: U.S. Congress, 1983. 407 pp.

This report is the result of a three-year effort, involving dozens of industrial, environmental, and policy experts. A thorough, well-organized document, it is the most authoritative government study of America's current predicament.

In addition to a readable 40-page summary of its findings, the book is divided into five chapters: an extensive chapter on available technologies, another on how to best manage risk, a third collecting the basic data, another reviewing current federal and state programs, and a fifth surveying various policy options.

The book fills a long-standing need for a basic reference text. Representative charts and passages are excerpted in appendix A.

OTA is a non-partisan analytic support agency that serves the U.S. Congress. Its purpose is to help Congress deal with the complex technical issues confronting society.

Copies of this OTA report are available at the U.S. Government Printing Office, Superintendent of Documents, Washington, DC 20402. The GPO stock number is 052–003–00901–3.

Selected Articles on Related Topics

ABSTRACTED FROM ENVIROLINE DATA BASE, ORGANIZED INTO TOPIC SECTIONS

SUBJECT: ALTERNATIVES TO LAND DISPOSAL

AU Campbell-Moni
TI Industrial Waste Reduction and Recovery
SO Alternatives, Winter 82, V10, N2-3, p59 (5)
YR 82
AB In recent years, many small businesses have begun to realize that reducing pollution can raise their company profits. In 1975, the 3M Company initiated a Pollution Prevention Pays program in an attempt to mitigate environmental costs during economic recession. The company's program was able to eliminate 75,000 tons of air pollution, 1,325 tons of water pollutants, 500 million gallons of polluted wastewater, and 2,900 tons of sludge/year; total savings for the U.S. facilities from these reductions resulted in a gain of $17.4 million in three years. Efforts by other companies to follow 3M's example are discussed. However, economic incentives to encourage waste reduction and recovery are still needed. (1 diagram, 1 drawing, 10 references, 1 table.)

AU Cleary-J-G., Ivins-D-D., Kehrberger-G-J., Ryan-C-P., Stuewe-C-W.
IN Hydroscience, Tennessee
TI NPDES Best Management Practices Guidance Document
SO NTIS Report PB80–135221, Dec. 79 (179)

YR 81
PT Special Report
AB Current practices used by industry to prevent the release of toxic and hazardous substances to receiving waters from non-point sources are reviewed. Available information on best management practices is evaluated for both baseline and advanced waste management concepts. A method of identifying best management practices based on pollutant sources and the chemical and physical properties of pollutant compounds is described.

AU Palmer-Paul
IN Zero Waste Systems, California
TI Recycling Hazardous Wastes: The Only Way to Go
SO CBE Env. Review, Oct.–Nov. 80, P13 (3)
YR 81
PT Survey Report
AB About 80 percent of the hazardous chemical wastes now being disposed of in landfills and other waste repositories could be recycled, saving both raw materials and money, and reducing the hazardous waste burden. The operations of Zero Waste Systems, Inc., a California company that recycles hazardous chemical wastes for reuse, are described. Samples of an industrial plant's waste streams are analyzed, and a recycling strategy for the constituents of the waste streams is proposed. Wastes are collected and processed, and recycled chemicals are marketed and delivered. (1 drawing, 2 photos.)

AU Senkan-Selim-M., Stauffer-Nancy-W.
IN MIT
TI What to Do with Hazardous Wastes
SO Technology Review, Nov.–Dec. 81, V84, N2, P34 (14)
YR 82
PT Feature Article
AB The disposal of hazardous chemical wastes is an old problem faced by industrialized civilization. The menace is evident: toxic wastes threaten our health and the environment. In the 1970s, a new public awareness of waste disposal problems led to outrage and terror at dumping methods. Several options for waste management are presented: (1) recycling within the industry that produces the waste; (2) selling to other industries; (3) treating and recycling within the same industry, selling to another industry; (4) disposing without pretreatment (an illegal practice when hazardous wastes are concerned). Preferable methods for dealing with the problem include source reduction to reduce the initial quantity of wastes produced, and waste recycling for reuse. (3 diagrams, 2 drawings, 3 tables.)

AU Taylor-Graham-C.
IN University of Denver
TI Socioeconomic Analysis of Hazardous Waste Management Alternatives

SO Presented at EPA Treatment of Hazardous Waste 6th Research Sym., Chicago, Mar. 17–20, 80, P132 (17)
YR 80
PT Technical Report
AB A methodology for analyzing the economic and social effects of alternative approaches to hazardous waste management is described. The methodology involves the generation of a series of environmental threat scenarios that might arise from the use of various waste management techniques. An interaction model links policy, decision-making, technological, and socioeconomic aspects of alternatives. (2 diagrams, 5 graphs, 11 references.)

AU Wassersug-Steve
IN EPA
TI A Future for Recycling—Federal Perspective
SO Recycling Today, Dec. 79, V17, N12, P93 (5)
YR 80
PT Feature Article
AB Experts estimate that increased efficiency and recycling could reduce U.S. energy consumption by about 25 percent. Recycling rates for paper, major metals, and glass for 1977 and 1979 are presented. Effects of the Resource Conservation and Recovery Act of 1976 (RCRA) on the recycling of wastes are discussed. The Act contains provisions for technical assistance; elimination of open dumping; research and development; a cabinet-level resource conservation committee; and federal procurement of recycled products. The Waste Exchange or Clearinghouse, a mechanism for promoting recovery of industrial wastes, is explained. Hazardous wastes are defined according to RCRA, and the regulations governing their control are reviewed. (1 photo, 1 table.)

SUBJECT: BUSINESS PERSPECTIVES

AU Bracken-Marilyn-C., Madati-P-J., Mercier-M., Ramani-R-V., Smeets-J., Hasa-Josef, Turner-L.
TI Toxic Chemicals and the Environment
SO Industry and Env., Oct.–Dec. 81, V4, N4 (18)
YR 82
PT Technical Feature
AB The benefits of the chemical age are undeniable, but the hazards of toxic substances bring with the benefits the need for legislative and regulatory control of chemical products. Topics covered include: U.S. toxic chemical control laws, the regulation and control of toxic substances in Czechoslovakia, and the labeling and classification of chemical substances in EEC. The U.N. International Program on Chemical Safety is reviewed, and procedures and plans for controlling toxic chemicals and substances in India and Tanzania are presented and evaluated. (4 diagrams, 1 photo, 2 tables.)

TI Hazardous Waste Managers Are Cleaning Up
SO Chemical Business, June 28, 82, P16 (4)

YR 82
PT Technical Feature
AB Treating and disposing of hazardous wastes is now a $400 million/year business, and it is expected to grow to an annual volume of around $1.5 billion by 1985. This does not include profits associated with solvent recycling, which accounted for $200 million in sales in 1981. The efforts of various hazardous waste management companies in handling such residues and in vying for business are surveyed. (4 photos.)

AU Krishnan-R., Hlavin-R.
IN Jacobs Engineering Co., California
TI Alternatives for Hazardous Waste Management in the Petroleum Refining Industry
SO NTIS Report PB81–113417, 1979 (245)
YR 82
PT Special Report
AB Alternatives to land disposal for the treatment and disposal of potentially hazardous wastes produced from petroleum refining are explored. Potential resource recovery and detoxification oriented techniques that can be employed in the treatment of these wastes are identified. These alternatives' operations, design, and possible means of implementation are described.

AU Newkumet-Chris
TI What Did RCRA Do for Solid Waste?
SO Solid Wastes Management, May 82, V25, N5, P70 (4)
YR 82
PT Technical Feature
AB The effects that the Resource Conservation and Recovery Act has had on solid wastes management are examined. The Resource Conservation and Recovery Act, despite its early promise, never achieved gains in municipal solid waste management. EPA began to phase down the program a few years ago by channeling money budgeted under the Act into hazardous waste management regulatory programs.

SUBJECT: EUROPEAN FACILITIES

AU Palmark-Mogens, Christensen-Axel-E., Rasmussen-Alfred
IN Chemcontrol A/S, Denmark
TI Chemcontrol Conducts Consultant Engineering Services concerning Waste Oil and Chemical Wastes
SO Chemcontrol A/S Consulting Services Report, June 80 (31)
YR 82

PT Special Report
AB Chemcontrol, a Danish consulting company conducting engineering services that cover systems and processes for the collection, transport, recycling, treatment, regeneration, and deposit of oily and chemical wastes, has developed a national toxic waste disposal program for Denmark. The uniqueness of the system lies in its ability to handle all types of chemical wastes from industrial, municipal, and agricultural sources, and the creation of one central processing and disposal point for the nation's wastes. The collection, transport, and disposal of waste oil and chemical wastes resulting from operational negligence or technical failure in industrial production are integral to a general abatement system developed by Chemcontrol. The background and services offered by the company are summarized, and the basic staffing and management of the control system are outlined and described. (12 diagrams, 4 drawings, 3 graphs, 6 photos.)

SUBJECT: GROUNDWATER CONTAMINATION

AN 82–01309. 8202
TI Groundwater Supplies: Are They Imperiled?
SO Conservation Foundation Letter, June 81 (8)
YR 82
PT Survey Report
AB The importance of groundwater as a crucial water resource will steadily increase as surface water supplies are depleted or contaminated. Contamination of groundwater resources is a major issue, and protection of these supplies is of vital importance. However, no comprehensive management systems for protecting these supplies from depletion or contamination exist. The current patchwork of laws is inadequate to deal with the complex matrix of groundwater problems. Proposed federal strategies for extending these resources and controlling water quality are surveyed. (1 photo, 28 references.)

SUBJECT: SITING HAZARDOUS WASTE FACILITIES

AU Anderson-Richard-F., Greenberg-Michael-R.
IN Rutgers University
TI Hazardous Waste Facility Siting: A Role for Planners
SO American Planning Assn. J., Spring 82, V48, N2, P204 (15)
YR 82
PT Technical Feature

AB Hazardous wastes present a critical environmental challenge for the U.S. in the 1980s. Types and volumes of hazardous wastes are reviewed, and their potential for adverse human health and ecological effects is surveyed. Research designed to assess land use suitability for siting new hazardous waste facilities is based on water supply issues and other physical and cultural siting criteria. Planners can play an important role by identifying suitable sites through their knowledge of land use and environmental regulation with the aid of the screening process described. (3 maps, 48 references, 4 tables.)

AU Berry-Jason
TI The World's Largest Hazardous Waste Treatment Plant: Is It Safe for Louisiana?
SO Dangerous Properties of Industrial Materials Report, Sep.–Oct. 81, V1, N7, P12 (11)
YR 82
PT Technical Feature
AB Hazardous waste poses a two-faceted problem of regulation and disposal to state and local governments. Industrial Tank Corp. of California (IT) began building the world's largest waste treatment facility in Louisiana in December 1980 as a response to this problem. Yet its plan has evoked controversy throughout the state. Unresolved questions pertaining to the plant's site, construction, and operation are being raised in the courts. The mushrooming of petrochemical plants along the Mississippi River into billion dollar industries has engendered some of the antagonism now being aimed at IT. (2 diagrams, 1 photo.)

TI Meeting the Challenges of Hazardous Waste in New York State
SO Public Policy Inst. of New York State Report, March 81 (37)
YR 81
PT Special Report
AB Key issues associated with the development of a comprehensive New York State hazardous waste disposal program are examined. A project is currently under way to create a high-technology waste facility that will focus on limited classes of waste that alternative techniques may not be able to handle. This facility would be privately owned and operated. Available methods to prioritize wastes sent to this facility are described. Techniques that will maximize the potential for successful siting of the facility are reviewed. A program to maximize waste recycling and reuse is presented. (1 diagram, 1 map, 47 references, 9 tables.)

AU Wolf-Sidney-M.
IN University of Indiana
TI Public Opposition to Hazardous Waste Sites
SO Boston College Env. Affairs Law Review, 1980, V8, N3, P463 (78)
YR 81

PT Survey Report

AB Public opposition to hazardous waste facilities is a reaction to the failure of federal and state governments in general, and the Resource Conservation and Recovery Act of 1976 in particular, to confront directly and effectively crucial hazardous waste issues. The history of hazardous waste pollution, and fast inadequate hazardous waste regulations are reviewed. Proposed EPA regulations do not provide adequate financial responsibility measures to ensure needed liability coverage for pollution at regulated waste sites or to ensure proper long-term care for abandoned waste sites. The major defect of national hazardous waste policy is its failure to reduce waste growth. (366 references.)

SUBJECT: STATE INITIATIVES

AU Curtis-Thomas-W.

TI The State of the States: Management of Environmental Programs in the 1980s.

SO U.S. National Governors Association Report, May 82 (13)

YR 82

PT Special Report

AB A survey on states' management of environmental programs, focusing on permit fees and other non-federal funding sources, is presented. The Reagan administration has proposed a 20 percent funding cutback in federal programs for fiscal year 83; the states are already adjusting to previous federal budget cuts and the continuing recession. The survey covers the following areas: State Air Programs Grants under Section 105 of the Clean Air Act, Water Quality Management Grants under Sections 106 and 205(G) of the Clean Water Act, Hazardous Waste Management Grants under Section 3011 of the Resource Conservation and Recovery Act, and Water Supply Management Grants under Section 1443 of the Safe Drinking Water Act. (9 tables.)

AU Finlayson-Robert-A.

TI Should State Rules Be Tougher Than EPA's?

SO Solid Wastes Management, May 82, V25, N5, P78 (5)

YR 82

PT Technical Feature

AB The effects that Reagan's new federalism will have on environmental programs is examined. Under the new federalism policy that has been drafted by the administration, many federal responsibilities will be handed back to the states. Findings of a recent Library of Congress study on the issue are cited. The study found that federal interest in achieving environmental goals may not be compatible with a shifting of responsibilities to the states. States with less environmental controls are likely to attract and keep industry, it was discovered.

AN 82-03832. 8207

AU Hartford-William-D.

TI Oregon's Hazardous Waste Management Status Report 1980
SO Oregon Department of Environmental Quality Report, Nov. 80 (62)
YR 82
PT Special Report
AB The history and policies of Oregon's Hazardous Waste Management Program are described. Industrial surveys show that about 675,000 cubic feet/year of hazardous waste is generated in the state; 55 percent comes from electronics assembly, metal and alloy manufacturing, and metal fabrication and machining. Some 52 of the 78 waste disposal site investigations completed do not reveal short- or long-term health or environmental hazards. Licensing and the "cradle-to-grave" management program for such wastes are reviewed. (3 diagrams, 1 map, 10 references, 23 tables.)

TI Hazardous Waste Management in Illinois
SO Illinois Inst. of Natural Resources Report 79/32, Oct. 79 (102)
YR 80
PT Special Report
AB The implications of the Resource Conservation and Recovery Act of 1976 for the generation, management, and disposal of hazardous wastes in Illinois are discussed. Current information on hazardous wastes in Illinois is inadequate; the state generates an estimated three million tons per year of hazardous waste, an amount that is expected to increase in the future but at an unknown rate. Most such waste is landfilled; some is lagooned, incinerated, or injected into deep wells. Illinois has 39 landfill sites that can accept hazardous waste, but some may reject the waste in the future, and no new sites are expected to become available. The economic impacts of the Act in Illinois are estimated at $4.2 million/year; disposal costs are more than three times greater than the federal estimate of disposal costs. Illinois costs are projected to rise 15 to 40 percent/year. (43 references, 20 tables.)

AU Marks-Bill
IN Michigan Department of Natural Resources
TI Managing Michigan's Hazardous Wastes
SO EPA Journal, June 81, V7, N6, P4 (3)
YR 81
PT Technical Feature
AB About 1.3 million tons of hazardous waste are created each year in Michigan. Most of the wastes have no permanent disposal area, and old wastes continue to plague the state with costly cleanups. The Hazardous Waste Management Act of 1979 gives local governments and the general public strong roles in selecting disposal sites and in choosing which disposal methods will be used. Local boards, responsible for reviewing and granting disposal site requests, have been appointed. The state plans to bring together all interests—federal, local, and state—into one broad and balanced waste disposal program. (2 photos.)

TI Technical, Marketing, and Financial Findings for the New York State Hazardous Waste Management Program
SO New York State Env. Facilities Corp./New York State Hazardous Waste Disposal Advisory Committee Report, March 80 (278)
YR 81
PT Special Report
AB The technical, marketing, and financial aspects of implementing a program for the environmentally sound treatment and disposal of non-radioactive hazardous wastes are assessed. Existing hazardous waste disposal practices are surveyed. Alternative hazardous waste treatment methods evaluated are storage; treatment, recycling, and disposal; waste exchange techniques; incineration; and secure landfilling. Until more extensive and conclusive data on the technical feasibility of the treatment methods are available, the state should not move toward implementation of a single long-range hazardous waste management plan. (Numerous diagrams, graphs, maps, references, tables.)

SUBJECT: SUMMARY OF REGULATIONS

AU Nalesnik-Richard-P., Kearney-Jerome-R.
IN Nalesnik Assoc., Washington, D.C.
TI An Analysis of Federal Legislative Jurisdictional Responsibilities for Toxic and Hazardous Materials
SO NTIS Report AD-A082 542, Feb. 80 (64)
YR 81
PT Special Report
AB Federal agency jurisdictional responsibilities for toxic substance and hazardous materials as presently authorized under existing legislation are delineated. Some 33 pieces of federal legislation are summarized, involving shared jurisdiction by 11 separate federal agencies. The implications of the complex network of legislative control of toxic substances and hazardous materials for quick and effective emergency management strategies are identified.

SUBJECT: TOWARD A NATIONAL TOXIC WASTE POLICY

AU Bronstein-Daniel A., Wennerberg-Linda-S.
IN Michigan State University
TI Section 8(B) of the Toxic Substances Control Act: A Case Study of Government Regulation of the Chemical Industry
SO Natural Resources Lawyer, 1981, V13, N4, P704 (24)
YR 82
PT Feature Article

AB Ways in which EPA approached the task of evaluating the chemical industry under the Toxic Substances Control Act are described. The first three years of work under the Control Act are discussed and the evolution of administrative rules is traced. Section 8(B) of the Control Act requires EPA to provide an inventory of chemicals that were produced and distributed three years before enactment. Chemicals not on that list are considered "new" and must pass EPA inspection before the go-ahead is given for production. Administrative problems faced by EPA in handling technical requirements are detailed. (134 references, 1 table.)

AU Ginsberg-William-R.
IN Hofstra University
TI Toward a National Hazardous Waste Policy
SO Dangerous Properties of Industrial Materials Report, May–June 82, V2, N3, P19 (4)
YR 82
PT Technical Feature
AB Problems with our national policy for hazardous wastes are identified. Federal efforts in this area can hardly be called "policy" due to the fragmentation and confusion that exists. A comprehensive public policy for hazardous wastes has been lacking primarily due to difficulties in governmental intervention with the production-consumption phenomenon.

TI Hazardous Waste Pollution: The Need for a Different Statutory Approach
SO Env. Law, Winter 82, V12, N2, P 443 (25)
YR 82
PT Survey Report
AB Shoddy industrial hazardous waste disposal practices and inadequate government monitoring have jeopardized groundwater supplies and public health. Regional hazardous waste disposals sites managed by quasi-government bodies are proposed. Common law remedies involve time-consuming burden of proof requirements while pollution continues unabated. Thus it is recommended that these remedies, available to victims of chemical contamination, should be replaced by a federal toxic contamination tort based on strict and vicarious liability with liberalized causation requirements. (140 references.)

AU Muskie-Edmund-S., Snyder-David-L.
IN Chadbourne Parke Whiteside & Wolff
TI Disposal, Energy Costs and Demands Spur Need for Systems
SO Solid Wastes Management, Oct. 81, V24, N10, P20 (4)
YR 82
PT Survey Report

AB The regulatory maze facing builders of waste-to-energy projects, which are expected to become economically attractive as landfilling costs soar, consists primarily of programs under the Clean Air Act and the Resource Conservation and Recovery Act. Federal and state regulations including new source review, best available control technology, lowest achievable emission rate, prevention of significant deterioration, emission offset requirements, OSHA-regulated pollutants, and hazardous waste regulations are explained. Sponsors of waste-to-energy systems should prepare an environmental compliance study at least two years before project construction commences.

Index

Contributors

JAMES BANKS is a senior staff attorney for the Natural Resources Defense Council. Since 1976, he has been director of NRDC's project on clean water and the principal litigant in numerous lawsuits involving the discharge of pollutants into waterways, including cases on EPA's pretreatment programs, the toxic discharge programs reported on in chapter 4, and EPA's consolidated permits regulations. Vice-Chairman of the Water Quality Committee from the American Bar Association, he holds a civil engineering degree from the University of Kansas (1972) and a law degree from the University of Michigan (1975).

KIRK W. BROWN is a professor at Texas A&M University. K. W. Brown & Associates (KWB&A) is a multidisciplinary staff of scientists and engineers who specialize in soil related aspects of the storage, treatment, and disposal of both hazardous and non-hazardous wastes. They also perform cleanup assessment of salt and chemical spills; compatibility testing of clay liner–waste combinations; reclamation of drastically disturbed lands; and interpretation of soil analyses. Throughout its research and consulting, KWB&A specializes in the movement and degradation of chemical compounds and plant nutrients in the soil environment. KWB&A clients have included petroleum refineries, chemical plants, waste disposal and mining companies, manufacturing facilities, other environmental consultants, law firms, public interest groups, individuals, and federal, state, and local government agencies.

SHEILA BROWN was counsel to the Energy and Commerce Committee of the U.S. House of Representatives from March 1979 to November of 1982. She drafted legislation on hazardous waste disposal, Superfund cleanup activities, nuclear waste transportation, and groundwater contamination and was also responsible for major amendments to the Resource Conservation and Recovery Act, the Hazardous Materials Transportation Act, and the Toxic Substances Control Act. Before joining the energy committee, she was a legislative counsel to Congressman John E. Moss of California and a staff member for Congressman

Jack Brooks. She received her BA in political science and history from Marquette University (1972) and her law degree from the McGeorge School of Law (1976).

GARY A. DAVIS was a hazardous waste management specialist with the California Governor's Office of Appropriate Technology from October 1980 to December 1982. He was a principal author of a major study on hazardous waste recycling and treatment technologies and policy initiatives to encourage their use. He was responsible for developing regulations to restrict the land disposal of certain hazardous wastes and also advised the California Department of Health Services on hazardous waste law and policy. Mr. Davis is now a practicing attorney in Knoxville, Tennessee, with the firm of Davis and Nickle, specializing in environmental law. He began his career as a chemical engineer with Hydroscience, Inc., in Knoxville, Tennessee, as a pollution control consultant for the chemical and pharmaceutical industries, in 1975. He earned a BS degree in chemical engineering from the University of Cincinnati (1975) and a JD from the University of Tennessee College of Law (1980).

GARY N. DIETRICH was director of the Office of Solid Waste of the Environmental Protection Agency from May 1981 to May 1982. Prior to that, from May 1978 to May 1981, he was deputy director of that same office. In those capacities, he was responsible for supervising the development of the hazardous waste management regulations under the Resource Conservation and Recovery Act. Mr. Dietrich culminated 23 years of federal service when he left EPA in May 1982 to become senior vice president for Clement Associates, an engineering and scientific consulting firm located in Arlington, Virginia. He started his career with the U.S. Public Health Service in 1957, moved to the Federal Water Quality Administration in 1966, and finally moved to EPA when that agency was formed in 1970. During his federal career he helped organize EPA, assisted in organizing the Office of Toxic Substances within EPA, helped develop the program to regulate PCBs, and managed EPA's program planning and budgeting process for several years, as well as the Superfund legislation and its ensuing program, to name a few of his major involvements. Mr. Dietrich earned a BA degree in civil engineering from the California Institute of Technology and completed graduate work at Cornell University. He is now president of Clement Associates.

JAMES FLORIO was first elected to the 94th Congress in 1974 and has been reelected to each successive Congress. A prime mover of efforts to establish the Superfund and to reauthorize the Resource Conservation and Recovery Act, Florio has taken a leading role in arguing for the closure of long-standing loopholes in federal regulations. He has been an active member of the Subcommittee on Health and the Environment and the Subcommittee on Public Lands and National Parks. He currently chairs the House Subcommittee on Commerce, Transportation, and Tourism. A graduate of Columbia University in public law and government, he received his law degree from Rutgers (1967).

RICHARD C. FORTUNA is Executive Director of the Hazardous Waste Treatment Council. Prior to his current position, he spent two years as staff toxicologist for the House Energy and Commerce Committee. From 1979-1981, he was legislative assistant to Congressman John Dingell. Throughout his career, Mr. Fortuna has helped to shape many of the amendments to the Resource Conservation and Recovery Act and the Superfund Act. He holds a bachelor of science degree in zoology and a masters of public health in toxicology from the University of Michigan in Ann Arbor.

KHRISTINE L. HALL is co-chairperson of and attorney for Environmental Defense Fund's toxic chemicals program. In that capacity, she works on legal and legislative approaches to hazardous waste, hazardous air pollution, and toxic chemicals generally. Ms. Hall was formerly a special projects attorney in the Land and Natural Resources Division of the Department of Justice and assistant director of the President's Task Force on Global Natural Resources and Environment. Prior to that, she worked as an attorney for environmental groups and taught courses in environmental law, natural resources law, and water law at the University of Kansas and University of Iowa. Ms. Hall received her juris doctor degree in 1974 and her bachelor of arts degree in slavic languages and literature in 1971, both from the University of Kansas.

JOEL HIRSCHHORN is a senior associate at the Office of Technology Assessment, the analytical support branch of the U.S. Congress that helps legislators anticipate the impacts of technological changes on policymaking. As project director of OTA's three-year study, *Technologies and Management Strategies for Hazardous Waste Control*, Dr. Hirschhorn has influenced a number of legislative efforts to modify and improve federal programs. He also directed OTA's study on the international competitiveness of the U.S. steel industry that is generally regarded as the premier government analysis of the problems facing the American steel industry. Prior to joining OTA, he was a full professor of engineering at the University of Wisconsin, Madison, author of a textbook on powder metallurgy, a management consultant, and a research metallurgist for Pratt and Whitney Aircraft. He received his Ph.D. in materials engineering from Rensselaer Polytechnic Institute and his masters and undergraduate degrees from the Polytechnic Institute of Brooklyn.

PATRICK McCANN is a senior associate for the Energy and Environment Division of Booz-Allen & Hamilton, Inc., a prestigious consulting firm. He has written some of the pivotal industry studies on America's hazardous waste management capabilities for the Environmental Protection Agency, including one of the first inventories of industry's treatment capacities. Recently an adviser for the President's Private Sector Survey, Mr. McCann received his degrees in economics.

PETER MONTAGUE is project administrator of the hazardous waste research program in the School of Engineering/Applied Science at Princeton University. He is the co-author of two books on toxic heavy metals, *Mercury* (Sierra Club, 1971) and *No World without End* (Putnam, 1976). For the last seven years, he has studied the land disposal of radioactive and chemical wastes, including a study of state-of-the-art landfills in New Jersey. Mr. Montague received his Ph.D. in American studies from the University of New Mexico in 1971.

ABOUT THE EDITOR

BRUCE PIASECKI teaches courses on environmental and energy issues at Clarkson University in Potsdam, New York. His work on alternatives to dumping hazardous wastes has appeared in numerous journals, including the *Washington Monthly* (January 1983) and *Science 83* magazine (September 1983). He is currently writing a cultural history of America in the age of environmentalism, *In Nature We Trust*, which analyzes recent problems in toxic waste management, enhanced oil production and pesticide manufacturing. He is also researching European hazardous waste systems with Gary A. Davis. He received his Ph.D. from Cornell University in 1981. Dr. Piasecki is also the founder and principal coordinator of the American Hazard Control Group, an advisory consulting group that provides legal, technical, policy, and business development advice on the best available means to treat and destroy toxic substances.